Insurance Systems in Times of Climate Change

Cornel Quinto

Insurance Systems in Times of Climate Change

Insurance of Buildings Against Natural Hazards

Springer

Schulthess

Cornel Quinto
Attorney at Law, LL.M.
Zurich
Switzerland

ISBN 978-3-642-22434-8 e-ISBN 978-3-642-22435-5
DOI 10.1007/978-3-642-22435-5
Springer Heidelberg Dordrecht London New York

Library of Congress Control Number: 2011939997

ISBN 978-3-7255-5891-9
www.schulthess.com

Originally published in German under the title: "Versicherungssysteme in Zeiten des Klimawandels. Elementarschaden-Versicherung von Gebäuden" © Schulthess Juristische Medien AG, Zurich – Basel – Geneva 2010

Translated by Traducta SA, Lausanne

Published by Springer-Verlag GmbH Berlin Heidelberg 2012

Printed on acid-free paper

Springer is part of Springer Science+Business Media (www.springer.com)

For Sabine

Preface

Climate change is one of the most pressing problems in the world of today—and of the future. Almost daily television images and press releases of devastating natural disasters are ominous. What awaits us? The first chapter provides an introductory overview, of the projected consequences of climate change and the institutional framework where the struggle for solutions is taking place.

Climate change affects all areas of life and strikes at the basis of our existence. This includes, last but not least, the "roof over one's head" and the workplace infrastructure—residential buildings, commercial and industrial properties. Protecting this basis of existence is of key significance, and this is bound to increase considerably in the course of climatic change. The present book deals with systems for protecting buildings against destruction and damage as a result of climate change. What demands does climate change make on a protection system that is fit for the future? Insurance against natural hazards is at the forefront of the solution. Climate change leads to new challenges and changes that insurance against natural hazards must meet if it is still to act as *insurance* and to grant insurance coverage under acceptable conditions to *all* building owners, as far as possible.

Already today, there is a multitude of existing systems for insuring buildings against natural hazards. The second chapter begins by illustrating the various types. Subsequently, the insurance systems in Germany, France, Spain, Switzerland and the USA are presented and compared with each other. The insurance system that is most suitable in view of climate change is determined and described. The third chapter draws a conclusion in this regard.

The legitimacy of the insurance systems is the subject of the fourth chapter. The focus is on making insurance statutory in the sense of compulsory insurance that is executed by a single insurance organisation furnished with a legal monopoly. In particular the monopoly is in a conflicting relationship with a free market economy and competition.

Keywords in the legal dispute are, on the one hand, economic freedom, the fundamental freedoms of the EU, and competition law. On the other hand, there is the common good, the public interest, the services of general (economic) interest and the universal service. Their compatibility with Swiss and EU law is examined, and solutions are illustrated. Finally, this book ends with a conclusion to these legal considerations.

Zurich, December 2009 Cornel Quinto

Table of contents

List of abbreviations

art.	article
BBl.	Swiss federal gazette (Bundesblatt)
BII	business interruption insurance
BJ	(Swiss) Federal Office of Justice ([Schweizerisches] Bundesamt für Justiz)
BV	Federal Constitution of the Swiss Confederation dated 18 April 1999 (SL 101)
CO	Federal law dated 30 March 1911 regarding the supplement to the Swiss Civil Code (fifth part: Swiss Code of Obligations) (SL 220)
CCR	Caisse Centrale de Réassurance
cf.	compare
CFR	Charter of Fundamental Rights of the European Union
COM	EU Commission
COP	Conference of the Parties to the UNFCCC
diss.	dissertation
E.	consideration
ECFI	European Court of First Instance
ECJ	European Court of Justice
ed., eds.	edition, editor(s)
e.g.	for example
ES pool	Swiss insurance pool for damage caused by natural hazards (Schweizerischer Elementarschaden-Pool)
et seq.	et sequens/et sequentes
etc.	et cetera
EU	European Union
FC	(Swiss) Federal Court (Schweizerisches Bundesgericht)
FCR	federal court ruling (of Swiss Federal Court)
FEMA	Federal Emergency Management Agency
FHCF	Florida Hurricane Catastrophe Fund
FMS	free movement of services
FOE	freedom of establishment
FS	Festschrift

GHG	greenhouse gases
GVG	building insurance law (Gebäudeversicherungsgesetz)
HC	hazard class
ICL	Federal law on insurance contracts (Insurance Contract Law) dated 2 April 1908 (SL 221.229.1)
i.e.	id est (in other words)
IPCC	Intergovernmental Panel on Climate Change
IRG	Interkantonale Risikogemeinschaft Elementar
IRV	Interkantonaler Rückversicherungsverband
KG	Federal law on cartels and other limitations of competition (antitrust law) dated 6 October 1995 (SL 251)
KGV	Kantonale Gebäudeversicherung(en) (Cantonal building insurance[s])
LLL	large loss limit
NFIP	National Flood Insurance Program
no.	number
NZZ	Neue Zürcher Zeitung
OcCC	Organe consultatif sur les changements climatiques
OJEU	Official Journal of the European Union
p.	page, pages
par.	paragraph
PML	Possible Maximum Loss
Pra	(Swiss) Federal court practice
SFC	(Swiss) Federal Council
SGEI	Services of general economic interest
SL	Systematic collection of (Swiss) federal legislation
Slg.	Official collection of the ECJ ("Sammlung")
SO	Ordinance on the supervision of private insurance companies (supervision ordinance/Aufsichtsverordnung) dated 9 November 2005 (SL 961.011)
SPM	Summary for policymakers
SUVA	Schweizerische Unfallversicherungsanstalt (Swiss accident insurance institution)
TEU	Treaty on European Union
TEC	Treaty Establishing the European Community

TFEU	Treaty on the Functioning of the European Union
UN	United Nations
UNFCCC	United Nations Framework Convention on Climate Change dated 9 May 1992
UNO	United Nations Organization
UVG	Federal law on accident insurance (accident insurance law) dated 20 March 1981 (SL 832.20)
VAG	Federal law regarding the supervision of insurance companies (Insurance Supervision Law/Versicherungsaufsichtsgesetz) dated 17 December 2004 (SL 961.01)
VKF	Vereinigung Kantonaler Feuerversicherungen (association of cantonal fire insurances)
vol.	volume
ZÜRS	Zonierungssystem für Überschwemmung, Rückstau und Starkregen (flood, backwater and heavy precipitation zoning system)

References

BIAGGINI GIOVANNI / LIENHARD ANDREAS / RICHLI PAUL / RICHLI FELIX, Wirtschaftsverwaltungsrecht des Bundes, 5th ed., Basel 2009 (cited BIAGGINI / LIENHARD / RICHLI / UHLMANN, Wirtschaftsverwaltungsrecht)

BORER JÜRG, Kartellgesetz, Zurich 2005 (cited BORER)

BRINKMANN HENRIK, Die Kantonalen Gebäudeversicherungen—Eine Standortbestimmung aus gesamtschweizerischer Sicht, Bern 1997 (cited BRINKMANN)

CALLIESS CHRISTIAN / RUFFERT MATTHIAS (eds.), EUV/EGV—Das Verfassungsrecht der Europäischen Union mit Europäischer Grundrechtscharta, Kommentar, 3rd ed., Munich 2007 (cited Author, in: CALLIES / RUPPERT)

CONSORCIO DE COMPENSACION DE SEGUROS, Natural Catastrophes Insurance Cover—A Diversity of Systems, Madrid 2008 (cited CONSORCIO, Diversity of Systems)

COTTIER THOMAS / OESCH MATTHIAS (eds.), Allgemeines Aussenwirtschafts- und Binnenmarktrecht [vol. XI of Schweizerisches Bundesverwaltungsrecht], Basel 2007 (cited Author, in: COTTIER / OESCH)

DAUSES MANFRED A. (ed.), Handbuch des EU-Wirtschaftsrechts, loose-leaf collection (as of March 2009), vol. 2, Munich (cited AUTHOR, in: DAUSES)

EHRENZELLER BERNHARD / MASTRONARDI PHILIPPE / SCHWEIZER RAINER J. / VALLENDER KLAUS A. (eds.), Die Schweizerische Bundesverfassung, Kommentar, 2nd ed., St. Gallen 2008 (cited AUTHOR, St. Galler Kommentar BV)

GLAUS URS / HONSELL HEINRICH (eds.), Gebäudeversicherung—Systematischer Kommentar, Basel 2009 (cited AUTHOR, Kommentar GV)

HÄFELIN ULRICH / MÜLLER GEORG / UHLMANN FELIX, Allgemeines Verwaltungsrecht, 5th ed., Zurich 2006 (cited HÄFELIN / MÜLLER / UHLMANN)

KERSCHNER FERDINAND (ed.), Handbuch Naturkatastrophenrecht, Vienna 2008 (cited AUTHOR, Handbuch Naturkatastrophenrecht)

KÜNZLER ADRIAN, Effizienz oder Wettbewerbsfreiheit?, Tübingen 2008, diss. University of Zurich 2008 (cited KÜNZLER)

OPPERMANN THOMAS / CLASSEN DIETER / NETTESHEIM MARTIN, Europarecht, 4th ed., Munich 2009 (cited OPPERMANN / CLASSEN / NETTESHEIM)

QUINTO CORNEL, Staatliche Versicherung gegen Elementarschäden in der EU und in der Schweiz—Vereinbarkeit mit dem EU-Recht, Bern 2009 (cited QUINTO)

RHINOW RENÉ / SCHEFER MARKUS, Schweizerisches Verfassungsrecht, Basel 2009 (cited RHINOW / SCHEFER)

RICHARDSON / GORDON / MOORE II (eds.), Natural Disaster Analysis after Hurricane Katrina—Risk Assessment, Economic Impacts and Social Implications, Cheltenham/UK and Northhampton/Massachusetts/USA 2008 (cited AUTHOR, in: RICHARDSON / GORDON / MOORE II)

RICHLI PAUL, Grundriss des schweizerischen Wirtschaftsverfassungsrechts, Bern 2007 (cited RICHLI, Wirtschaftsverfassungsrecht)

SANDMANN TINA, Kommunale Unternehmen im Spannungsfeld von Daseinsvorsorge und europäischem Wettbewerbsrecht, Schriften zum Europarecht und zur Rechtsvergleichung, vol. 16, Frankfurt a. M. 2005, diss. University of Frankfurt a. M. 2004 (cited SANDMANN)

SCHWARZE JÜRGEN (ed.), EU-Kommentar, 2nd ed., Baden-Baden 2009 (cited AUTHOR, in: Schwarze)

SCHWARZE REIMUND / WAGNER GERT G., Naturgefahren in Europa—Vielfältige Antworten auf den Klimawandel, in: SCHÄFER WOLF / SCHNEIDER ANDREA / THOMAS TOBIAS, Märkte und Politik—Einsichten aus Perspektive der Politischen Ökonomie, Marburg 2009, p. 175 et seq. (cited SCHWARZE / WAGNER, in: Märkte und Politik)

SPINNER BRUNO / MARITZ DANIEL, EG-Kompatibilität des schweizerischen Wirtschaftsrechts: Vom autonomen zum systematischen Nachvollzug, in: FORSTMOSER PETER / VON DER CRONE HANS CASPAR / WEBER ROLF H. / ZOBL DIETER (eds.), Der Einfluss des europäischen Rechts auf die Schweiz—Festschrift für Roger Zäch zum 60. Geburtstag, Zurich 1999, p. 127 et seq. (cited SPINNER / MARITZ, in: FS Zäch)

STREIFF MATTHIAS, Von der formellen und materiellen Beendigung des Rückversicherungsvertrages, Zurich 2000, diss. University of Zurich 1999 (cited STREIFF)

SWISS RE, Reinsurance matters—A manual of the non-life branches, Zurich 2005 (cited SWISS RE, Manual)

TERCIER PIERRE / BOVET CHRISTIAN (eds.), Commentaire Romand, Droit de la concurrence, Geneva/Basel/Munich 2002 (cited AUTHOR, Commentaire Romand LCart)

THÜRER DANIEL / AUBERT JEAN FRANÇOIS / MÜLLER PAUL (eds.), Verfassungsrecht der Schweiz, Zurich 2001 (cited AUTHOR, Verfassungsrecht der Schweiz)

TSCHANNEN PIERRE / ZIMMERLI ULRICH / MÜLLER MARKUS, Allgemeines Verwaltungsrecht, 3rd ed., Bern 2009 (cited TSCHANNEN / ZIMMERLI / MÜLLER)

UEBERSAX PETER, Privatisierung der Verwaltung, ZBl 2001, p. 393 et seq. (cited UEBERSAX, ZBl 2001)

UHLMANN FELIX, Gewinnorientiertes Staatshandeln, Basler Studien zur Rechtswissenschaft, series B Öffentliches Recht, vol. 52, Basel 1997, diss. University of Basel 1996 (cited UHLMANN)

VALLENDER KLAUS A. / HETTICH PETER / LEHNE JENS, Wirtschaftsfreiheit und begrenzte Staatsverantwortung, 4th ed., Bern 2006 (cited VALLENDER / HETTICH / LEHNE)

VERSICHERUNGSWIRTSCHAFT, Risiko Hochwasser: Per Mausklick die richtige Gefahrenzone, 4/2000, p. 275 (cited VW 2000, Hochwasser)

VOGEL STEFAN, Der Staat als Marktteilnehmer, diss. University of Zurich 2000 (cited VOGEL, Der Staat als Marktteilnehmer)

VON BÜREN ROLAND / DAVID LUCAS (eds.), Schweizerisches Immaterialgüter- und Wettbewerbsrecht, vol. V/2 Kartellrecht, Basel 2000 (cited AUTHOR, SIWR V/2).

VON FÜRSTENWERTH FRANK / WEISS ALFONS, Versicherungsalphabet, 9th ed., Karlsruhe 1997 (cited VON FÜRSTENWERTH / WEISS)

VON UNGERN-STERNBERG THOMAS, Gebäudeversicherung in Europa—Die Grenzen des Wettbewerbs, Bern 2002 (cited VON UNGERN-STERNBERG)

WANNER CHRISTINE, 100 Jahre zeitgemäss—Meilensteine in der Brand- und Elementarschaden-Versicherung in der Schweiz, Bern 2002 (cited WANNER).

WEBER ROLF H. / UMBACH PATRICK, Versicherungsaufsichtsrecht, Bern 2006 (cited WEBER / UMBACH)

ZÄCH ROGER, Schweizerisches Kartellrecht, 2nd ed., Bern 2005 (cited ZÄCH, Kartellrecht)

ZÄCH ROGER / KÜNZLER ADRIAN, Traditionelle europäische Wettbewerbspolitik versus "more economic approach", EuZ 209, p. 30 et seq. (cited ZÄCH / KÜNZLER, EuZ 2009)

List of materials

Alternative Finanzierungs- und Versicherungslösungen, Vergleich unterschiedlicher Risikotransfersysteme dreier vom Augusthochwasser 2005 betroffenen Länder: Deutschland, Österreich und die Schweiz *[Alternative financing and insurance solutions, comparison of different risk transfer systems in three countries affected by flooding in August 2005: Germany, Austria and Switzerland],* Präventionsstiftung der Kantonalen Gebäudeversicherungen (ed.), Paul Raschky / Reimund Schwarze / Manijeh Schwindt / Hannelore Weck-Hannemann (authors), Bern 2009 (cited Alternative Financing and Insurance Solutions)

Bericht des Bundesrates "Grundversorgung in der Infrastruktur (Service public)" *[Report by the federal council "Basic service provision in infrastructure (public services)"]* dated 23 June 2004, BBl 2004, 4569 et seq. (cited Federal Council Report "Basic Service Provision in Infrastructure")

Bericht "Klimaänderung und die Schweiz 2050—Erwartete Auswirkungen auf die Umwelt, Gesellschaft und Wirtschaft" *[Report "Climate change and Switzerland 2050—Anticipated effects on the environment, society and economy"]* ed. OcCC/Pro Clim., March 2007 (cited Report "Climate Change and Switzerland 2050")

Botschaft des BR zur Änderung des Bundesgesetzes über die Unfallversicherung *[Communication by the SFC on the amendment to the federal law on accident insurance]* dated 29 May 2008, BBl 2008, 5395 et seq. (cited UVG Communication)

Botschaft des BR zum Nationalen Finanzausgleich *[Communication by the SFC on national financial adjustment]*, BBl 2002, 2291 et seq. (cited NFA Communication)

Botschaft des BR zum Abkommen zwischen der Schweiz und der Europäischen Wirtschaftsgemeinschaft betreffend die Direktversicherung mit Ausnahme der Lebensversicherung und zum Bundesgesetz über die Direktversicherung mit Ausnahme der Lebensversicherung (Schadensversicherungsgesetz [SchVG]) *[Communication by the SFC on the Agreement between the Swiss Confederation and the European Economic Community concerning direct insurance other than the life insurance and the federal law on direct insurance other than the life insurance (indemnity insurance act)]* dated 14 August 1991, BBl 1991 IV 1 et seq. (cited Direct Insurance Agreement and SchVG Communication)

Botschaft des Bundesrates betreffend den Entwurf eines Gesetz über den Geschäftsbetrieb von Privatunternehmungen im Gebiete des Versicherungswesens *[Communication by the SFC regarding the draft law on the business of private undertakings in the insurance sector]* dated 13 January 1885, BBl 1885 I 101 et seq. (cited VAG Communication 1885)

Botschaft des BR zu einem neuen Bundesgesetz über die Beaufsichtigung privater Versicherungseinrichtungen *[Communication by the SFC regarding a new federal law on the supervision of private insurance organisations]* dated 5 May 1976, BBl 1976 II 873 et seq. (cited VAG Communication 1978)

Botschaft des BR zu einem Gesetz betreffend die Aufsicht über Versicherungsunternehmen (Versicherungsaufsichtsgesetz, VAG) und zur Änderung des Bundesgesetzes über den Versicherungsvertrag *[Communication by the SFC on a law regarding the supervision of insurance companies (Insurance Supervision Law, VAG) and on an amendment to the federal law on insurance contracts]* dated 9 May 2003, BBl 2003, 3789 et seq. (cited VAG Communication 2006)

CCR, Les Catastrophes Naturelles en France/Natural disasters in France, September 2008 (cited CCR, Catastrophes Naturelles)

FEMA, Prepared. Responsive. Committed., FEMA B-653, July 2008 (cited FEMA, Prepared. Responsive. Committed.)

First Assessment Report of the IPCC dated 30 August, 1990 (cited IPCC AR1)

Gesamtverband der Deutschen Versicherungswirtschaft (GDV), Land unter—Schutz vor Überschwemmungen und Hochwasser (Land under—protection from floods and deluges), Berlin 2009 (brochure)

Gesamtverband der Deutschen Versicherungswirtschaft (GDV), Stürmische Zeiten – Schäden vorbeugen und richtig versichern (Stormy times—preventing damage and properly insuring), Berlin 2009 (brochure)

Gutachten des Bundesamtes für Justiz vom 29. September 2009 betreffend die Revision UVG zuhanden der SGKN *[Expert report by the federal office of justice dated 29 September 2009 regarding the revision of the UVG for the attention of the SGKN]* (cited Expert Report BJ on UVG revision)

Interkantonaler Rückversicherungsverband (IRV) annual report 1995

Interkantonaler Rückversicherungsverband (IRV) annual report 1998

Interkantonaler Rückversicherungsverband (IRV) annual report 1999

Interkantonaler Rückversicherungsverband (IRV) annual report 2004

Interkantonaler Rückversicherungsverband (IRV) annual report 2006

Interkantonaler Rückversicherungsverband (IRV) annual report 2008

Interkantonaler Rückversicherungsverband (IRV) bylaws dated 23 June 1982 (as of 2009) (cited IRV bylaws)

Leitfaden des Kantons St. Gallen "Naturgefahren im Kanton St. Gallen, Leitfaden für Vorsorge und Schutz" *[Guideline by the Canton of St. Gallen "Natural hazards in the Canton of St. Gallen, Guideline for prevention and protection"]*, Naturgefahren-komission des Kantons St. Gallen, 2007 (cited Natural Hazards Guideline St. Gallen)

Communication from the Commission "Services of general interest, including social services of general interest: a new European commitment", COM (2007) 725 final, dated 20 November 2007 (cited COM Communication "Services of general interest")

Communication from the Commission "White paper on services of general interest", COM (2004) 374 final, dated 12 May 2004 (cited COM "White Paper Services of general interest")

Communication from the Commission "Services of general interest in Europe", COM (2000) 0580 final, dated 20 September 2000, OJEU C017 dated 19 January 2001, p. 4 et seq. (cited COM Communication "Services of General interest in Europe")

Commission interpretative communication "Freedom to provide services and the general good in the insurance sector" dated 16 February 2000, OJEU C043 dated 16 February 2000, p. 5 et seq. (cited "Insurance Sector Interpretative Communication")

Münchener Rück/Munich Re, Sonderheft Millenium "topics 2000—Naturkatastrophen, Stand der Dinge" *[Natural disasters—state of affairs]*, Munich, December 1999, with insert by NatCat Service, Münchener Rück/Munich Re, REF/Geo, January 2000 (cited Munich Re, topics 2000)

Münchener Rück/Munich Re, Topics Geo, Natural catastrophes 2008—*Analyses, assessments, positions*, Munich 2009 (cited Munich Re, topics 2008)

Muster Elementarschaden-Rückversicherungsvertrag IRV *[model natural hazards reinsurance contract]* (as of 2009)

National Flood Insurance Program, Myths and Facts about the National Flood Insurance Program, FEMA, F-002 (03/07) (cited FEMA leaflet, Myths and Facts about the National Flood Insurance Program)

Principles Governing IPCC Work, approved at the Fourteenth Session, Vienna, 1 October 1998 (cited Principles Governing IPCC Work)

Rapport Annuel Caisse Centrale de Réassurance (CCR) 2005

Rapport Annuel Caisse Centrale de Réassurance (CCR) 2007

Rapport Annuel Caisse Centrale de Réassurance (CCR) 2008

Summary for Policymakers, approved in Valencia, 12–17 November 2007, part of the IPCC Fourth Assessment Report (AR4), Climate Change 2007, Synthesis Report (SYR) (cited IPCC SPM 2007)

Swiss Re, Folgen der Klimaveränderung: Mehr Sturmschäden in Europa *[Consequences of climate change: more storm damage in Europe]*, Fokus Report, Zurich 2006 (cited Swiss Re, Storm losses)

Swiss Re, Überschwemmungen sind versicherbar! *[Floods are insurable!]*, Focus Report, Zurich 2002 (cited Swiss Re, Floods)

Swiss Re, Chancen und Risiken der Klimaänderung *[Chances and risks of climate change]*, Zurich 2002 (cited Swiss Re, Climate Change)

Synthesis Report of the congress on "Climate Change—Global Risks, Challenges & Decisions" dated 10–12 March 2009, University of Copenhagen

UN General Assembly Resolution 43/53 dated 6 December 1988

Vereinigung Kantonaler Feuerversicherungen (VKF) annual report 1998

Vereinigung Kantonaler Feuerversicherungen (VKF) annual report 1999

Vereinigung Kantonaler Feuerversicherungen (VKF) annual report 2000

Vereinigung Kantonaler Feuerversicherungen (VKF) annual report 2005

Vereinigung Kantonaler Feuerversicherungen (VKF) annual report 2006

Vereinigung Kantonaler Feuerversicherungen (VKF) annual report 2007

Wegleitung Objektschutz gegen gravitative Naturgefahren *[Guidance on protecting property against gravitational natural hazards]*, Thomas Egli (author), VKF (ed.), Bern 2005 (cited Guidance on Protecting Property Gravitational Natural Hazards)

Wegleitung Objektschutz gegen meteorologische Naturgefahren *[Guidance on protecting property against meteorological natural hazards]*, Thomas Egli (author), VKF (ed.), Bern 2007 (cited Guidance on Protecting Property Meteorological Natural Hazards)

Chapter 1: Challenge: Climate change as the starting situation

I. Global climate change

1. IPCC—global reference for a global problem

Climate change affects all parts of the world, all continents. Some countries may even benefit from climate change in the coming years in certain areas (e.g. higher agricultural yields in Northern countries). Other areas, however, are developing unfavourably (water shortage, melting glaciers) and the "bottom line" is that the effects, which are noticeable even now, will be distinctly negative (see p. 7 et seq. below). Thus, the effects are not limited to certain parts of the world. It is, in the true sense of the word, a global problem.

Initially, however, the issue was, and is, to recognise and to acknowledge climate change as a global problem. The latter in particular called for a globally accepted reference. Such a reference must have a broad basis and must be objective and politically neutral, otherwise it will not command the required approval, thus remaining ineffective. The institutional framework of this reference will be described briefly below.

It is not a coincidence that this global reference—the Intergovernmental Panel on Climate Change (IPCC)—was created within the framework of the UNO as a world-spanning organisation. The IPCC was founded in 1988 by the World Meteorological Organization (WMO) as well as the United Nations Environment Programme (UNEP). It is a joint sub-organisation of the UN organisations WMO and UNEP. Membership is open to all Member States of the WMO or UNEP[1].

The IPCC's task consists of comprehensively, objectively, openly and transparently evaluating the scientific, technical and socio-economical information that is pertinent to the scientific understanding of climate change caused by man. This is intended to make an objective source of information on the causes of climate change and its effects available to politics. The IPCC reports to that effect are expected to deal with politically relevant issues (e.g. water shortage, potential measures to reduce climate change) but should not attempt to dictate the political route to be taken. The IPCC should not itself "make" policy but remain politically neutral and supply politics with the information required for political decision making[2].

---

1 Principles Governing IPCC Work, no. 7; UN General Assembly Resolution 43/53 dated 6 December 1988.

2 Cf. Principles Governing IPCC Work, no. 2, as well as the large volume of information at www.ipcc.ch.

The IPCC received the Nobel Peace Prize in 2007. Its reports are met with broad approval and have achieved a situation whereby the international community is now taking the first steps to confront climate change effectively. The secret of success, apart from the above-mentioned principles, lies in the fact that the authorship of the IPCC reports has a very broad political and geographical basis and that the reports are subjected to a sophisticated review process before they are definitively approved[3]. Furthermore, the IPCC is an intergovernmental organisation; reports are approved by representatives of the Member States and the content is thereby recognised by these states. The results of the IPCC's work are thus based on a broad consensus both in form as well as in content.

2. The IPCC as the initial spark for the UNFCCC and the Kyoto Protocol

The UN General Assembly took the initiative in its 44th session in 1989 that had led to the founding of the IPCC. It held out the prospect of adopting a resolution regarding the start of negotiations on a framework agreement to fight climate change by its 45th session in 1990 after the IPCC had submitted its first report. In 1990, the IPCC completed its "First Assessment Report"[4]. The UN General Assembly picked up the impetus and decided in its 45th session to effectively start negotiations for a framework agreement with the objective of completing it by 1992, on the occasion of the scheduled UN Conference on Environment and Development (UNCED) in Rio ("Rio Earth Summit"). The undertaking was successful. The United Nations Framework Convention on Climate Change (UNFCCC) was adopted on 9 May 1992 within the framework of the UNCED and entered into force on 21 March 1994.

Although the UNFCCC does not contain any specific, mandatory measures to fight climate change, its purpose consists of establishing a framework for future efforts to fight climate change. Its merit, however, is to have set the goal of stabilising the concentration of greenhouse gases (GHG) at a level that prevents the dangerous interference with the climate system caused by man (anthropogenic) (albeit without specific and mandatory objectives or even reduction measures). In addition, the industrialised countries among the Member States are obligated to keep an ongoing inventory of GHG emissions, and all Member States commit to developing national programmes to curb climate change (art. 2, art. 4 no. 1 (a) and (b) UNFCCC).

---

3 The three working groups of the IPCC (see footnote 10 below) are each presided over by a co-chairman from an industrialised nation and a co-chairman from an emerging or a developing nation. The reports are examined by experts and government representatives in each case and reviewed and edited accordingly before being approved. The "Summary for Policymakers" (SPM), the most-respected and most-read report by the IPCC, is even discussed line by line in the IPCC plenary session and approved by the same (cf. Principles Governing IPCC Work, Appendix A).

4 IPCC AR1 dated 30 August 1990.

In addition, the UNFCCC has paved the way for the Kyoto Protocol, which is based on art. 2 and 3 UNFCCC[5]. In contrast to the UNFCCC, the Kyoto Protocol contains specific objectives that must be met by 2012, namely, among other things, the reduction of GHG emissions in the industrialised countries and the EU states by 5% compared to the level of 1990 (art. 3 of the Kyoto Protocol).

Since the first reduction period of the Kyoto Protocol ends in 2012, new, specific objectives were negotiated intensively within the framework of the regular conferences of the contracting states of the UNFCCC, the COP ("Conferences of the Parties to the UNFCCC") and corresponding preparatory rounds of negotiations (e.g. the "Bonn Climate Change Talks")[6]. The preliminary climax of this process was COP 15 which is commonly titled "climate summit". It took place in Copenhagen from 7–19 December 2009.

3. The results of the Copenhagen "climate summit"

Not only the contracting states of the UNFCC, including states that have not yet ratified the Kyoto Protocol (in particular the USA), convened at the COP 15, but also representatives of numerous non-governmental organisations. A total of 193 states were represented which makes the COP 15 one of the largest UNO conferences, with respect to participation, ever to have taken place.

As measured by the expectations and especially in view of the urgency of the problem, the results of the COP 15 turned out to be modest. It became apparent even before the start of the conference that it would not result in the conclusion of a binding successor protocol that would replace the Kyoto Protocol. All the same, all states, including the most important GHG emitters such as the USA, China and India, were able to agree on a joint convention, the "Copenhagen Accord". The wording is printed in the footnote[7]. In view of the contents, one cannot but agree with the state-

---

[5] The Kyoto Protocol was adopted on 11 December 1997 and entered into force on 16 February 2005.

[6] Information on the negotiation process is posted and updated continuously at www.unfccc.int, with numerous documents available for downloading.

[7] Decision -/CP.15
The Conference of the Parties,
Takes note of the Copenhagen Accord of 18 December 2009.
Copenhagen Accord
The Heads of State, Heads of Government, Ministers, and other heads of the following delegations present a the United Nations Climate Change Conference 2009 in Copenhagen: [List of Parties]
In pursuit of the ultimate objective of the Convention as stated in its Article 2,
Being guided by the principles and provisions of the Convention,
Noting the results of work done by the two Ad hoc Working Groups,
Endorsing decision x/CP.15 on the Ad hoc Working Group on Long-term Cooperative Action and decision x/CMP.5 that requests the Ad hoc Working Group on Further Commitments of Annex I Parties under the Kyoto Protocol to continue its work,
Have agreed on this Copenhagen Accord which is operational immediately.

1. We underline that climate change is one of the greatest challenges of our time. We emphasise our strong political will to urgently combat climate change in accordance with the principle of common but differentiated responsibilities and respective capabilities. To achieve the ultimate objective of the Convention to stabilize greenhouse gas concentration in the atmosphere at a level that would prevent dangerous anthropogenic interference with the climate system, we shall, recognizing the scientific view that the increase in global temperature should be below 2 degrees Celsius, on the basis of equity and in the context of sustainable development, enhance our long-term cooperative action to combat climate change. We recognize the critical impacts of climate change and the potential impacts of response measures on countries particularly vulnerable to its adverse effects and stress the need to establish a comprehensive adaption programme including international support.
2. We agree that deep cuts in global emissions are required according to science, and as documented by the IPCC Fourth Assessment Report with a view to reduce global emissions so as to hold the increase in global temperature below 2 degrees Celsius, and take action to meet this objective consistent with science and on the basis of equity. We should cooperate in achieving the peaking of global and national emissions as soon as possible, recognizing that the time frame for peaking will be longer in developing countries and bearing in mind that social and economic development and poverty eradication are the first and overriding priorities of developing countries and that a low-emission development strategy is indispensable to sustainable development.
3. Adaption to the adverse effects of climate change and the potential impacts of response measures is a challenge faced by all countries. Enhanced action and international cooperation on adaption is urgently required to ensure the implementation of the Convention by enabling and supporting the implementation of adaption actions aimed at reducing vulnerability and building resilience in developing countries, especially in those that are particularly vulnerable, especially least developed countries, small island developing States and Africa. We agree that developed countries shall provide adequate, predictable and sustainable financial resources, technology and capacity-building to support the implementation of adaption action in developing countries.
4. Annex 1 Parties commit to implement individually or jointly the quantified economy-wide emissions targets for 2010, to be submitted in the format given in Appendix I by Annex I Parties to the secretariat by 31 January 2010 for compilation in an INF document. Annex I Parties that are Party to the Kyoto Protocol will thereby further strengthen the emissions reductions initiated by the Kyoto Protocol. Delivery of reductions and financing by developed countries will be measured, reported and verified in accordance with existing and any further guidelines adopted by the Conference of the Parties, and will ensure that accounting of such targets and finance is rigorous, robust and transparent.
5. Non-Annex I Parties to the Convention will implement mitigation actions, including those to be submitted to the secretariat by non-Annex I parties in the format given in Appendix II by 31 January 2010, for compilation in an INF document, consistent with Article 4.1 and Article 4.7 and in the context of sustainable development. Least developed countries and small island developing States may undertake actions voluntarily and on the basis of support. Mitigation actions subsequently taken and envisaged by Non-Annex I Parties, including national inventory reports, shall be communicated through national communications consistent with Article 12.1(b) every two years on the basis of guidelines to be adopted by the Conference of the Parties. Those mitigation actions in national communications or otherwise communicated to the Secretariat will be added to the list in appendix II. Mitigation actions taken by Non-Annex I Parties will be subject to their domestic measurement, reporting and verification the result of which will be reported through their national communications every two years. Non-Annex I Parties will communicate information on the implementation of their actions through National Communications, with provisions for international consultations and analysis under clearly defined guidelines that will ensure that national sovereignty is respected. Nationally appropriate mitigation actions seeking international support will be recorded in a registry along with relevant technology, finance and

capacity building support. Those actions supported will be added to the list in appendix II. These supported nationally appropriate mitigation actions will be subject to international measurement, reporting and verification in accordance with guidelines adopted by the Conference of the Parties.

6. We recognize the crucial role of reducing emissions from deforestation and forest degradation and the need to enhance removals of greenhouse gas emissions by forests and agree on the need to provide positive incentives to such actions through the immediate establishment of a mechanism including REDD-plus, to enable the mobilization of financial resources from developed countries.
7. We decide to pursue various approaches, including opportunities to use markets, to enhance the cost-effectiveness of, and to promote mitigation actions. Developing countries, especially those with low emitting economies should be provided incentives to continue to develop on a low emission pathway.
8. Scaled up, new and additional, predictable and adequate funding as well as improved access shall be provided to developing countries, in accordance with the relevant provisions of the Convention, to enable and support enhanced action on mitigation, including substantial finance to reduce emissions from deforestation and forest degradation (REDD-plus), adaption, technology development and transfer and capacity-building, for enhanced implementation of the Convention. The collective commitment by developed countries is to provide new and additional resources, including forestry and investments through international institutions, approaching USD 30 billion for the period 2010–2012 with balanced allocation between adaption and mitigation. Funding for adaption will be prioritized for the most vulnerable developing countries, such as the least developed countries, small island developing States and Africa. In the context of meaningful mitigation actions and transparency on implementation, developed countries commit to a goal of mobilizing jointly USD 100 billion dollars a year by 2020 to address the needs of developing countries. This funding will come from a variety of sources, public and private, bilateral and multilateral, including alternative sources of finance. New multilateral funding for adaption will be delivered through effective and efficient fund arrangements, with a governance structure providing for equal representation of developed and developing countries. A significant portion of such funding should flow through the Copenhagen Green Climate Fund.
9. To this end, a High Level Panel will be established under the guidance of and accountable to the Conference of the Parties to study the contribution of the potential sources of revenue, including alternative sources of finance, towards meeting this goal.
10. We decide that the Copenhagen Green Climate Fund shall be established as an operating entity of the financial mechanism of the Convention to support projects, programme, policies and other activities in developing countries related to mitigation including REDD-plus, adaption, capacity-building, technology development and transfer.
11. In order to enhance action on development and transfer of technology we decide to establish a Technology Mechanism to accelerate technology development and transfer in support of action on adaption and mitigation that will be guided by a country-driven approach and be based on national circumstances and priorities.
12. We call for an assessment of the implementation of this Accord to be completed by 2015, including in light of the Convention's ultimate objective. This would include consideration of strengthening the long-term goal referencing various matters presented by the science, including in relation to temperature rises of 1.5 degrees Celsius.

ment by UN Secretary General Ban Ki-moon: *"This accord cannot be everything that everyone hoped for, but it is an essential beginning"*[8]. *The* Copenhagen Accord merely marks the beginning of the efforts to negotiate a workable climate protection agreement. The most important points of the former are listed below:

- The Copenhagen Accord does not contain any binding, specific reduction goals with respect to GHG emissions. The industrialised countries have merely committed to implement those reduction goals that they had specified as of the end of January 2010 by 2020. It is not defined how high the reduction has to be. All the same, the contracting states of the Kyoto Protocol have committed to going beyond the scope of the reduction goals of the protocol. Developing countries are merely obliged to implement the mitigation actions specified by the end of January 2010 (cf. no. 4 and 5 Copenhagen Accord). The media have therefore criticised the Copenhagen Accord as a "non-binding accord" or a mere "declaration of intent"[9].
- Industrialised countries have committed to have their specific reductions verified by an international, still to be refined, mechanism. Developing and emerging nations, however, do not have to submit to international verification but will merely have to verify their mitigation actions by themselves; this was designed to accommodate China's demand, which rejected any international verification for reasons of sovereignty. Even so, mitigation actions that receive international (especially financial) support, must be amenable to international verification (cf. no. 4 and 5 Copenhagen Accord).
- It is recognised that the increase in temperature must be kept below 2 °C in order to prevent "dangerous" anthropogenic climate change (cf. no. 1 and 2 Copenhagen Accord). No. 12 then states that strengthening the objectives, including in relation to a rise in temperature of (only) 1.5 °C should be considered in 2015. This probably takes into account the fact that the IPCC and the sciences, as shown below, contrary to the "2-degree goal" constantly propagated by politics, do not just assume climate change to be "dangerous" only at a rise in temperature of 2 °C. Instead, the IPCC, depending on the global region and exposure, assumes that even significantly lower warming has severe consequences. The 2 °C are a political compromise, not a scientific guideline (see p. 13 et seq. below).

---

[8] UNO, UNFCCC, Secretariat, Press release, Copenhagen, 19 December 2009. Numerous documents of the COP 15, in particular various decisions and the working papers of various working groups and boards, as well as the Copenhagen Accord, can be downloaded from www.unfccc.int.

[9] Cf. NZZ no. 296 of 21 December 2009, p. 3, "Ein schwacher Rahmen für die Klimapolitik" (*A weak framework for climate policy*); Tages-Anzeiger of 21 December 2009, p. 7, "Der Durchbruch, der keiner war" (*The breakthrough that wasn't one*).

- The considerable problem of forest logging and its negative effects on the climate is expressly recognised and immediate remedial action, including through financial means, is promised (cf. no. 6 Copenhagen Accord).
- To fight climate change and to finance measures to that effect, including in particular by developing countries, the industrialised countries commit to making USD 30 billion available for the period from 2010 to 2012. Starting in 2020, the industrialised countries should, in addition, make USD 100 billion available each year in support of the developing countries (cf. no. 8 Copenhagen Accord).
- An international mechanism shall be created to promote technology transfer with respect to technologies that curtail climate change or allow adjustment (cf. no. 11 Copenhagen Accord).
- A "Copenhagen Green Climate Fund" shall be created that, among other things, implements the financing of projects and technology transfer (cf. no. 8 and 10 Copenhagen Accord).

The conference participants have not specified a target date for concluding a binding succession protocol after Kyoto or any other deadline for the arrangement of effective and extensive reduction goals. This, too, was and is (vehemently) criticised.

The COP 16 shall take place in Cancun (Mexico) from 29 November–10 December 2010. A preceding round of negotiations in Bonn from 31 May–11 June 2010 was also agreed. It remains to be seen whether the community of states will achieve the objective of counteracting climate change by taking decisive measures.

4. What awaits us worldwide

A. The prognoses

By now, it is undisputed that climate change is already taking place and is caused by man, particularly the sharp rise in GHG emissions (especially CO_2) caused by man (anthropogenic) . All crucial indicators, namely

- surface temperature
- sea level
- glacier and ice coverage (polar ice caps)
- snow coverage
- amount of precipitation
- number and magnitude of extreme weather events

show changes that are obvious the world over and that prove climate change in the sense of global climate warming and that cannot be explained by natural climate fluctuations[10]

Even with respect to the forecasts up to 2100, there is a high probability that these are accurate. How the above-mentioned indicators will change in the future, will depend substantially on the development of future GHG emissions. The IPCC has developed a variety of emissions scenarios[11] where it is unavoidable in each case that GHG emissions will continue to rise for the time being as long as only the current climate change mitigation is being implemented. Even in the best case, GHG emissions (unless significant reductions gain ground starting immediately) will not reduce until 2050. Compared to the 1980–1999 level, the following forecast results:

- Surface temperature by 2100: even in the best case[12], + 1.8 °C, with a spread of + 1.1 to + 2.9 °C
- Sea level by 2100: even in the best case, +0.18 to +0.38 m

The following figure from the IPCC SPM 2007 (cf. footnote 10) shows the GHG emissions scenarios from 2000 onward on the left and the rise in temperature compared to the period from 1980 to 1999 on the right (in English according to the source, IPCC)[13]:

10 Cf. p. 2 et seq. and p. 5 et seq. IPCC SPM 2007. Each IPCC assessment report consists of the reports of the individual working groups, with regard to the IPCC Fourth Assessment Report (AR4) of 2007, of the reports "The Physical Science Basis" (Working Group I), "Impacts Adaption and Vulnerability" (Working Group II) and "Mitigation of Climate Change" (Working Group III). In addition, each assessment report contains a synthesis report (SYR) that comprises a summary of the working group reports and any additional special reports, if applicable. Finally, the "Summary for Policymakers" (SPM) constitutes a summary of the summary that illustrates the most important findings of the SYR again in condensed form. The present findings come from the IPCC SPM 2007 that was adopted in Valencia on 12–17 November 2007 (cited IPCC SPM 2007). All reports can be downloaded from www.ipcc.ch.

11 Scenarios according to IPCC Special Report on Emissions Scenarios (SRES, 2000).

12 B1 scenario.

13 Source: IPCC SPM 2007, p. 7, Figure SPM.5.

Scenarios for GHG emissions from 2000 to 2100 (in the absence of additional climate policies) and projections of surface temperatures

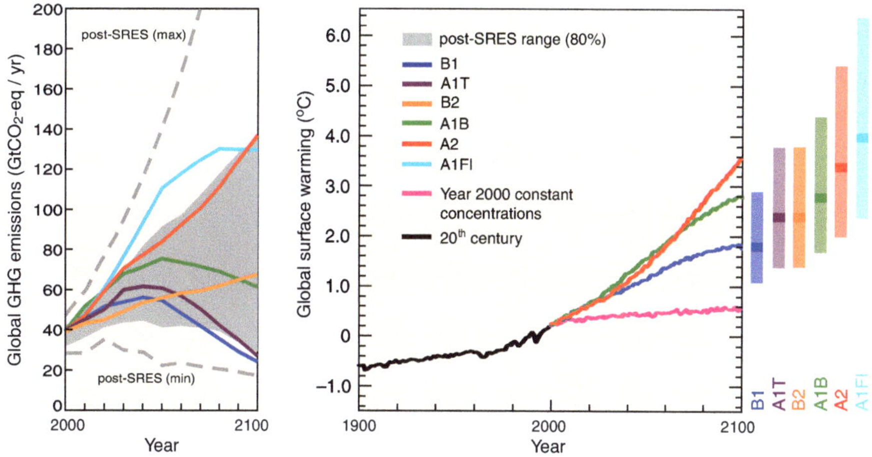

With an unchecked further increase in GHG emissions (scenarios A2 and A1Fl), a 3.4–4 °C increase in temperature as well as a sea level rise of 0.26–0.59 m can even be expected. Such changes would overtax the adaption capacity both of man and of the ecosystem and would have devastating consequences (see p. 11 et seq. below).

The following table from the IPCC SPM 2007 shows the forecast rise in surface temperature and sea levels according to the various scenarios compared to the 1980–1999 level[14].

Case	Temperature change (°C at 2090-2099 relative to 1980-1999) [a, d]		Sea level rise (m at 2090-2099 relative to 1980-1999)
	Best estimate	*Likely* range	Model-based range excluding future rapid dynamical changes in ice flow
Constant year 2000 concentrations[b]	0.6	0.3 – 0.9	Not available
B1 scenario	1.8	1.1 – 2.9	0.18 – 0.38
A1T scenario	2.4	1.4 – 3.8	0.20 – 0.45
B2 scenario	2.4	1.4 – 3.8	0.20 – 0.43
A1B scenario	2.8	1.7 – 4.4	0.21 – 0.48
A2 scenario	3.4	2.0 – 5.4	0.23 – 0.51
A1FI scenario	4.0	2.4 – 6.4	0.26 – 0.59

Notes:

a) Temperatures are assessed best estimates and *likely* uncertainty ranges from a hierarchy of models of varying complexity as well as observational constraints.

b) Year 2000 constant composition is derived from Atmosphere-Ocean General Circulation Models (AOGCMs) only.

c) All scenarios above are six SRES marker scenarios. Approximate CO_2-eq concentrations corresponding to the computed radiative forcing due to anthropogenic GHGs and aerosols in 2100 (see p. 823 of the Working Group I TAR) for the SRES B1, AIT, B2, A1B, A2 and A1FI illustrative marker scenarios are about 600, 700, 800, 850, 1250 and 1550ppm, respectively.

d) Temperature changes are expressed as the difference from the period 1980-1999. To express the change relative to the period 1850-1899 add 0.5°C.

These forecasts do not take into account the negative feedback effects that arise due to the fact that as the result of warming a) forests and oceans can absorb even less

[14] Source: IPCC SPM 2007, p. 8, Table SPM.1.

CO_2 as well as that b) the polar ice caps are not able to reflect as much sunlight, which in turn accelerates warming[15].

The following figure from the IPCC SPM 2007 shows the forecast rise in surface temperature for various parts of the world until 2100 compared to 1980–1999, namely for the middle scenario A1B[16]:

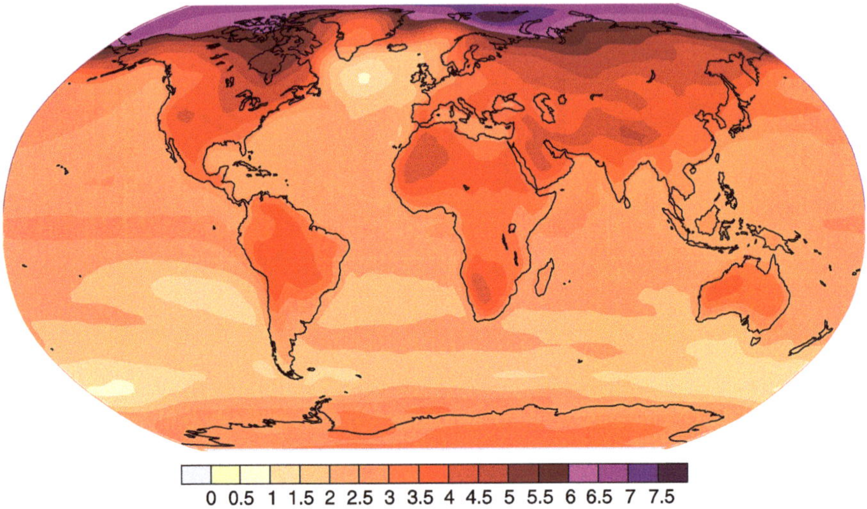

According to this scenario, a 3 °C rise in temperature is forecast for Central Europe while a rise of more than 4 °C is expected for large parts of Africa.

In addition, the following consequences regarding climate and weather events are expected in all probability due to the ongoing warming[17] (see pp. 16 et seq. below for Switzerland specifically):

- Thawing of permafrost
- Increase in the intensity of tropical storms
- Increase in heavy precipitation
- Increase in flooding and inundation as the result of storms, heavy precipitation and rise in sea levels

The following table by the IPCC SPM 2007 provides an overview of the phenomena to be expected after 2050–2100, their likelihood of occurrence and the forecast so-

[15] IPCC SPM 2007, p. 7 et seq.; cf. also p. 15 et seq. below and footnote 30.

[16] Source: IPCC SPM 2007, p. 9, Figure SPM.6.

[17] IPCC SPM 2007, p. 8 and 13.

cio-economic impacts. With respect to heavy precipitation events, tropical cyclones and extreme high sea level, the loss of property and the discontinuation of insurance coverage provided by private insurers are forecast[18]:

Phenomenon[a] and direction of trend	Likelihood of future trends based on projections for 21st century using SRES scenarios	Examples of major projected impacts by sector			
		Agriculture, forestry and ecosystems	Water resources	Human health	Industry, settlement and society
Over most land areas, warmer and fewer cold days and nights, warmer and more frequent hot days and nights	*Virtually certain*[b]	Increased yields in colder environments; decreased yields in warmer environments; increased insect outbreaks	Effects on water resources relying on snowmelt; effects on some water supplies	Reduced human mortality from decreased cold exposure	Reduced energy demand for heating; increased demand for cooling; declining air quality in cities; reduced disruption to transport due to snow, ice; effects on winter tourism
Warm spells/heat waves. Frequency increases over most land areas	*Very likely*	Reduced yields in warmer regions due to heat stress; increased danger of wildfire	Increased water demand; water quality problems, e.g. algal blooms	Increased risk of heat-related mortality, especially for the elderly, chronically sick, very young and socially isolated	Reduction in quality of life for people in warm areas without appropriate housing; impacts on the elderly, very young and poor
Heavy precipitation events. Frequency increases over most areas	*Very likely*	Damage to crops; soil erosion, inability to cultivate land due to waterlogging of soils	Adverse effects on quality of surface and groundwater; contamination of water supply; water scarcity may be relieved	Increased risk of deaths, injuries and infectious, respiratory and skin diseases	Disruption of settlements, commerce, transport and societies due to flooding: pressures on urban and rural infrastructures; loss of property
Area affected by drought increases	*Likely*	Land degradation; lower yields/crop damage and failure; increased livestock deaths; increased risk of wildfire	More widespread water stress	Increased risk of food and water shortage; increased risk of malnutrition; increased risk of water- and food-borne diseases	Water shortage for settlements, industry and societies; reduced hydropower generation potentials; potential for population migration
Intense tropical cyclone activity increases	*Likely*	Damage to crops; windthrow (uprooting) of trees; damage to coral reefs	Power outages causing disruption of public water supply	Increased risk of deaths, injuries, water- and food-borne diseases; post-traumatic stress disorders	Disruption by flood and high winds; withdrawal of risk coverage in vulnerable areas by private insurers; potential for population migrations; loss of property
Increased incidence of extreme high sea level (excludes tsunamis)[c]	*Likely*[d]	Salinisation of irrigation water, estuaries and fresh-water systems	Decreased fresh-water availability due to saltwater intrusion	Increased risk of deaths and injuries by drowning in floods; migration-related health effects	Costs of coastal protection versus costs of land-use relocation; potential for movement of populations and infrastructure; also see tropical cyclones above

Notes:
a) See Working Group I Table 3.7 for further details regarding definitions.
b) Warming of the most extreme days and nights each year.
c) Extreme high sea level depends on average sea level and on regional weather systems. It is defined as the highest 1% of hourly values of observed sea level at a station for a given reference period.
d) In all scenarios, the projected global average sea level at 2100 is higher than in the reference period. The effect of changes in regional weather systems on sea level extremes has not been assessed.

The following figure shows the intensity of the socio-economic impacts in relation to the rise in temperature. While certain impacts are forecast even for a rise in temperature of less than 1 °C, the considerable flooding of coast lines is expected, for example, only after a rise in temperature of at least 2 °C. The bottom figure ("Warming by 2090–2099 relative to 1980–1999 for non-mitigation scenarios"), however, shows that a rise in temperature of less than 2 °C can be expected only in the best case (scenario B1), while middle scenarios (B2, A1B) assume a rise of more than 2 °C already[19].

[18] Source: IPCC SPM 2007, p. 13, Table SPM.3.
[19] Source: IPCC SPM 2007, p. 10, Figure SPM.7.

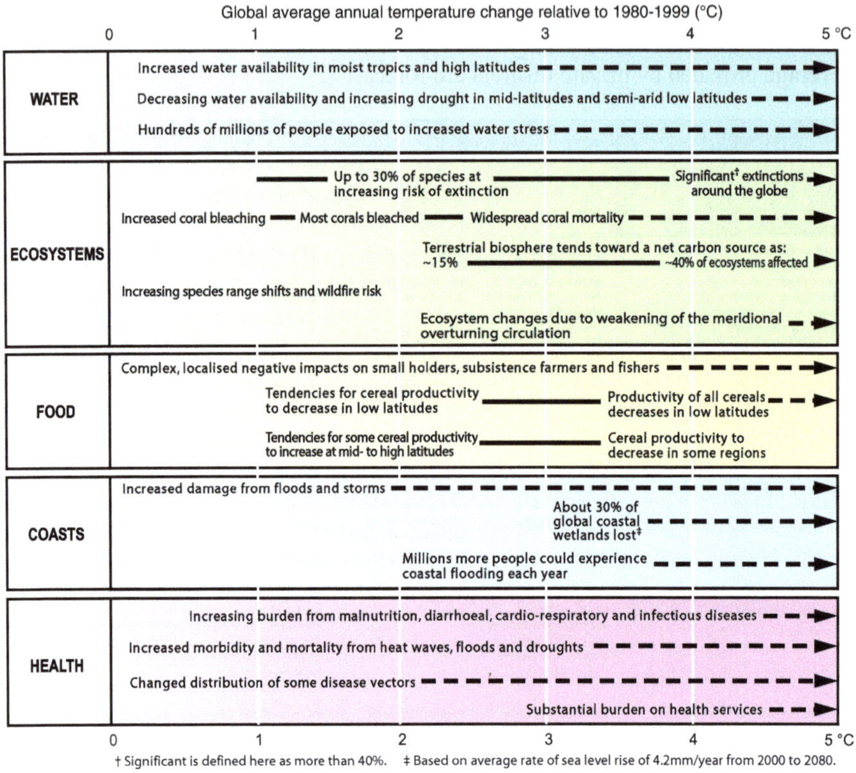

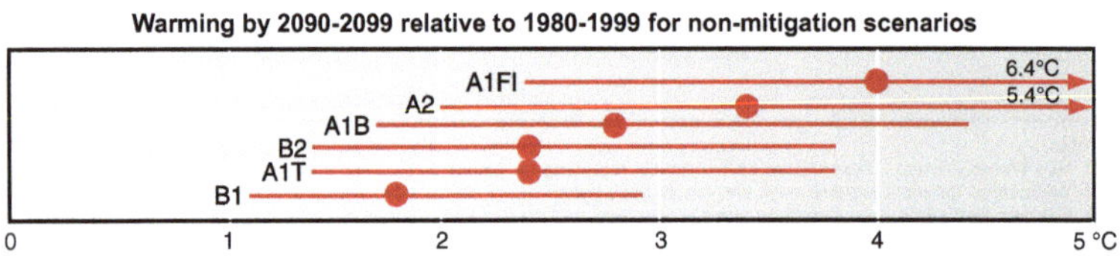

In addition, there is the danger of abrupt as well as irreversible impacts of climate change. Thus, it cannot be ruled out that over a temporal horizon of one hundred or several hundred years, the partial loss of the polar ice caps may result in sea levels rising by several metres; this would lead to floods in low-lying areas and to considerable changes in coast lines. Furthermore, it is likely that, with a rise in surface temperature by more than 1.5–2.5 °C (compared to the level of 1980–1999) about 30% of all species will become extinct[20].

20 IPCC SPM 2007, p. 13.

B. Adaption to and mitigation of climate change as a way out?

Adaption measures aim at adapting to climate change and absorbing the negative effects as well as possible. The chances for success of this strategy depend not least on the adaption capacity, which exists to the extent only in which suitable technologies, financial means and socially stable conditions exist. Nevertheless, even societies with high adaption capacity, namely affluent western societies, are vulnerable to climate change[21].

Mitigation actions, by contrast, are to curb climate change itself and its negative effects and prevent them, if possible. The implementation of this strategy calls for, among other things, an immediate significant reduction in GHG emissions. This can be accomplished, to a much greater extent than with adaption, only by the massive use of state-of-the art technologies to that effect.

Adaption involves, for example, the use of new plant varieties (in particular those that need less water), the treatment of rain water or, as regards settlement, the creation of adapted insurance options. Mitigation actions primarily aim to drastically reduce energy consumption in all sectors (buildings, traffic, industry etc.), in particular by reducing the use of fossil fuels, the application of energy-efficient technologies and substitution by solar and other alternative energies[22].

There is a consensus that neither adaption nor mitigation measures alone can curb the risks of climate change. Instead, both types of measures must be taken[23].

When determining specific reduction goals, the objective is also not least to agree on when (in particular at what increase in surface temperature) a "dangerous" anthropogenic climate change is on hand. Such is the case if serious effects are expected in the most diverse of areas that considerably affect the ecosystem and human civilisation and threaten our existence to that effect, e.g. the loss of coastal areas due to flooding. The IPCC assumes that a "dangerous" climate change exists even if temperatures rise by 1–2 °C above the level of 1990 or by 1.5–2.5 °C above the *pre-industrial level* (1750)[24]. For the time being, the EU has committed itself politically to the theory that a rise in temperature of over 2 °C above the *pre-industrial level* will trigger a "dangerous" climate change. The July 2009 G8 summit in L'Aquila has recognised this limit[25].

---

21 IPCC SPM 2007, p. 14.

22 Cf. tables with potential adaption and mitigation measures, IPCC SPM 2007, p. 15 and 17.

23 IPCC SPM 2007, p. 19.

24 IPCC SPM 2007, p. 19.

25 Council of the European Union, 2005: Presidency Conclusions, Brussels 22/23 March 2005, EU Commission, Brussels. In the closing communiqué of the Major Economies Forum on Energy and Climate, 9 July 2009, the G-8 have agreed on the *"recognition of the scientific assessment that global warming must not exceed 2 degrees Celsius compared to the pre-industrial level (translation, original quote in German)"*. NZZ of 10 July 2009, p. 1.

Since the climate system reacts only very sluggishly, sea levels and surface temperatures will continue to rise significantly over the coming decades even if emissions are reduced considerably. In the best case, a stabilisation of the GHG concentration at a lower level will be possible around the year 2020. Limiting the rise in temperature to 2–2.4 °C and the sea level rise to 0.4–1.4 m compared to the pre-industrial level also requires a reduction of CO_2 emissions by 85–50% compared to the emissions level in the year 2000, and this by no later than 2050[26]. To achieve this, however, GHG emissions must be *massively reduced all over the world starting immediately* according to this. The following table by the IPCC shows a variety of GHG stabilisation scenarios, what level of CO_2 reduction they require, and what rise in temperature and sea levels is to be expected in this respect[27].

Category	CO_2 concentration at stabilisation (2005 = 379 ppm)[b]	CO_2-equivalent concentration at stabilisation including GHGs and aerosols (2005 = 375 ppm)[b]	Peaking year for CO_2 emissions[a,c]	Change in global CO_2 emissions in 2050 (percent of 2000 emissions)[a,c]	Global average temperature increase above pre-industrial at equilibrium, using 'best estimate' climate sensitivity[d,e]	Global average sea level rise above pre-industrial at equilibrium from thermal expansion only[f]	Number of assessed scenarios
	ppm	ppm	year	percent	°C	metres	
I	350 – 400	445 – 490	2000 – 2015	-85 to -50	2.0 – 2.4	0.4 – 1.4	6
II	400 – 440	490 – 535	2000 – 2020	-60 to -30	2.4 – 2.8	0.5 – 1.7	18
III	440 – 485	535 – 590	2010 – 2030	-30 to +5	2.8 – 3.2	0.6 – 1.9	21
IV	485 – 570	590 – 710	2020 – 2060	+10 to +60	3.2 – 4.0	0.6 – 2.4	118
V	570 – 660	710 – 855	2050 – 2080	+25 to +85	4.0 – 4.9	0.8 – 2.9	9
VI	660 – 790	855 – 1130	2060 – 2090	+90 to +140	4.9 – 6.1	1.0 – 3.7	5

Notes:

a) The emission reductions to meet a particular stabilisation level reported in the mitigation studies assessed here might be underestimated due to missing carbon cycle feedbacks (see also Topic 2.3).

b) Atmospheric CO_2 concentrations were 379ppm in 2005. The best estimate of total CO_2-eq concentration in 2005 for all long-lived GHGs is about 455ppm, while the corresponding value including the net effect of all anthropogenic forcing agents is 375ppm CO_2-eq.

c) Ranges correspond to the 15th to 85th percentile of the post-TAR scenario distribution. CO_2 emissions are shown so multi-gas scenarios can be compared with CO_2-only scenarios (see Figure SPM.3).

d) The best estimate of climate sensitivity is 3°C.

e) Note that global average temperature at equilibrium is different from expected global average temperature at the time of stabilisation of GHG concentrations due to the inertia of the climate system. For the majority of scenarios assessed, stabilisation of GHG concentrations occurs between 2100 and 2150 (see also Footnote 21).

f) Equilibrium sea level rise is for the contribution from ocean thermal expansion only and does not reach equilibrium for at least many centuries. These values have been estimated using relatively simple climate models (one low-resolution AOGCM and several EMICs based on the best estimate of 3°C climate sensitivity) and do not include contributions from melting ice sheets, glaciers and ice caps. Long-term thermal expansion is projected to result in 0.2 to 0.6m per degree Celsius of global average warming above pre-industrial. (AOGCM refers to Atmosphere-Ocean General Circulation Model and EMICs to Earth System Models of Intermediate Complexity.)

The following figure from the IPCC shows once again that a limitation of the temperature rise to around 2 °C can be achieved only with the most favourable stabilisation scenario (Scenario I, green). All other scenarios assume a temperature rise of significantly more than 2 °C[28]. Apart from that, it must be emphasised that the IPCC and the sciences do not classify climate change as being "dangerous" starting from a 2 °C rise in temperature only, as it is disseminated in politics time and time again. Numerous serious effects are to be expected, depending on region and exposure, at a far lower increase already.

---

26 IPCC SPM 2007, p. 20 et seq.

27 Source: IPCC SPM 2007, p. 20, Table SPM.6.

28 Source: IPCC SPM 2007, p. 21, Figure SPM.11.

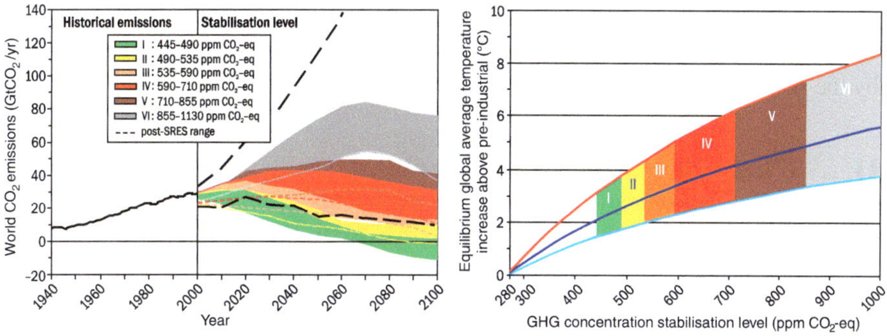

Recently, a synthesis report that was prepared as the outcome of a meeting on climate change in Copenhagen from 10–12 March 2009 caused a stir[29]. The report is not a document issued by the IPCC but numerous renowned scientists have contributed to it. It scared the interested public, for the scientists assume an acceleration of climate change based on the latest climate data that had been gathered after 2006 and that are not yet incorporated into the most recent Assessment Report by the IPCC of 2007, the AR4 (cf. footnote 10). It shows that GHG emissions are already developing in the direction of the IPCC's most pessimistic emissions scenarios and that even the climate indicators are found in the uppermost range of the IPCC prognoses, or even exceed them. Thus, a rise in sea levels twice that of the 2007 IPCC prognoses is forecast, namely a probable rise of 1 m during the course of the 21st century (compared to 0.38 m in the best case or 0.59 m according to the scenario considered to be the likely as per IPC AR4). In addition, a rise in surface temperature in the uppermost range of the IPCC prognoses is forecast. These alarming assumptions are, among other things, ascribed to the melting of the polar ice caps, which is taking place significantly more extensively and faster both with respect to scope as well as time (facilitated, among other things, by feedback effects)[30]. Furthermore, based on the current development of the GHG concentration in the atmosphere, it is assumed that temperatures will probably rise by 2.0–2.4 °C compared to pre-industrial levels, which would mean that the 2 °C limit of "dangerous" climate change can no longer be adhered to[31]. Altogether, therefore, a significant acceleration and worsening of

[29] Synthesis Report of the "Climate Change—Global Risks, Challenges & Decisions" congress from 10–12 March 2009, University of Copenhagen (www.climatecongress.ku.dk). The synthesis report was presented by the authors in Brussels on 18 June 2009 and handed to the Danish Prime Minister Rasmussen, the host of the COP 15 in December 2009 in Copenhagen.

[30] Ice and snow reflect solar radiation and thereby absorb less warmth. Water, however, absorbs warmth from solar radiation. When ice recedes, more warmth is absorbed by the ice-free water which in turn continuously accelerates the melting process of the ice.

[31] Cf. in particular Synthesis Report by the above-named congress, chapter "Climatic Trends" and "Long-Term Strategy: Global Targets and Timetables" ("Key Message" 1 and 3). Cf. also NZZ no. 143 dated 24 June 2009, p. 7, "Ein Klimabericht mit politischer Note" (A climate report with a political emphasis).

climate change compared to the 2007 IPCC prognoses is forecast. It remains to be seen to what extent these assumptions are sustained in the next Assessment Report by the IPCC.

II. Climate change in Switzerland

1. The prognoses for Switzerland until 2050

The Consulting Organ for issues of climate change (OcCC)[32], by order of the Federation, has examined the consequences of climate change for Switzerland until 2050 and has outlined them in a report[33]. The IPCC Assessment Reports were used as the scientific basis (in particular the AR4), based on which a regional climate scenario was prepared, with corresponding prognoses. On this basis, the specific anticipated effects until 2050 were subsequently outlined as well.

The year 2050 was chosen as the temporal horizon because, on the one hand, climate scenarios by the IPCC still lie relatively close together up to this point in time; this allows effects to be outlined without having to differentiate between different scenarios. On the other hand, 2050 is a temporal horizon that still lies within the lifetime of today's generations, whereby today's generations feel that this concerns them as well, thus making them aware of their responsibilities (cf. Report "Climate Change and Switzerland 2050", p. 9).

Specifically, the following, among other things, is forecast until 2050; the changes are going to appear insidiously:

- Surface temperature (average) in winter + 1.8, in summer + 2.7 °C
- Increase in extreme weather events in general
- Increase in heavy precipitation, generally greater fluctuations and greater intensity of precipitation
- More severe westerly storm events

32 The OcCC (the abbreviation comes from Organe consultatif sur les changements climatiques) was founded in 1996 by the Swiss Federal Department of Home Affairs (EDI) and of Environment/Transport/Energy/Communications (UVEK) and is tasked with formulating recommendations on issues of climate and climate change for the attention of politics and administration.

33 Report "Klimaänderung und die Schweiz 2050—Erwartete Auswirkungen auf Umwelt, Gesellschaft und Wirtschaft" (Climate change and Switzerland 2050—Anticipated effects on environment, society and economy), OcCC/Pro Clim, editor, March 2007 (accessible at www.occc.ch or www.proclim.ch, cited Report "Climate Change and Switzerland 2050").

- Melting of the glaciers (smaller and medium-sized ones are going to disappear completely by 2050); according to a study by ETH Zurich, these have lost 12% of their volume in the last ten years alone[34]
- Thawing of permafrost
- Increase in heat waves and droughts
- Overall decrease in the amount of precipitation throughout the year in spite of heavy precipitation events, with more rain (instead of snow) falling in winter

2. The ramifications

The above prognoses have diverse ramifications for human living conditions, the economy and the environment. Only a few of the points of the above-named report are singled out below that are pertinent to insurance, in particular of buildings, against natural hazards:

- Increase in floods
- Increase in rock avalanches
- Increase in debris flows and landslides
- Overall increase in natural hazards
- Overall increase in the risk of damage for buildings (settlements) and infrastructure

In addition, numerous further effects are predicted in a wide range of sectors, only a few of which are listed below:

- Water shortages (as a result of droughts, decrease in overall amount of precipitation and loss of water stored in glacier ice, which will also lead to a drop in the groundwater level)
- Decline in energy production due to the decrease in running water (hydraulic power stations) and cooling water (nuclear power plants)
- Increase in the competition for water among consumers, agriculture (increasing need for irrigation) and energy producers
- Ecosystems such as forests and low-moor bogs come under increasing pressure, extinction of species
- Settlements, infrastructure and transport routes are exposed to increased risks and must be protected by preventive measures (with respect to insurance, see

[34] The study was published in the June 2009 issue of the magazine "Global and Planetary Change", published by Elsevier, Amsterdam. Cf. NZZ no. 142 dated 23 June 2009, p. 11.

p. 21 et seq. below; with respect to constructional measures, see pp. 28, 58 et seq. below; furthermore, spatial planning measures are required)

- Diverting to crops (agriculture) that are adapted to the climate will be necessary
- Negative impact on human health and productive capacity (heat waves)
- Other countries are going to experience even more serious effects of climate change compared to Switzerland. This in turn is going to have negative consequences for the Swiss economy and society. Trade relationships and raw material supply will be negatively affected (export, dependence on raw material). Massive population migrations are predicted, which will lead to social upheavals.

This listing could be continued at will and shows that climate change will have repercussions in countless sectors of society, economy and environment, and negative ones at that, as a rule.

III. Consequences for insurance against natural hazards

1. Effects on insurance against natural hazards

A. Climate change—no surprise for the insurance sector

Based on its loss statistics and loss trends, insurance companies providing insurance against natural hazards (including reinsurances) have been assuming for years already that a climate change was in progress. Insurances are not political players, allowing them to deal with the issue in a businesslike manner, free from the attachments to political positions in the matter of climate change.

The Munich Re, for example, picked up on the subject of climate change as long ago as 1973, in is special publication "Überschwemmungen" ("Floods"), as can be seen from its special edition "topics 2000—Naturkatastrophen—Stand der Dinge" (Natural disasters—state of affairs). In the latter, it published several lists on the development of natural catastrophes and their associated national economic and insured losses, including loss trends, for the period of 1950–1999, from a global perspective. In addition, the question of by which factor the number of natural disasters and the amount of loss have changed was examined in a decade-by-decade comparison and comparing, for example, the 1960s with the 1990s. The corresponding figures are replicated below[35]:

[35] Source: Münchener Rück/Munich Re, special issue Millenium "topics 2000—Naturkatastrophen, Stand der Dinge" (Natural disasters—state of affairs), Munich, December 1999, p. 104, with insert by the NatCatService, Münchener Rück/Munich Re, REF/Geo, January 2000 (cited Munich Re, topics 2000).

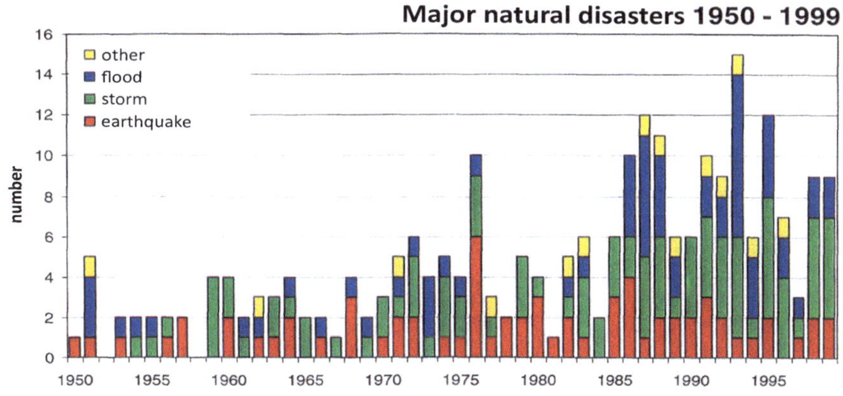

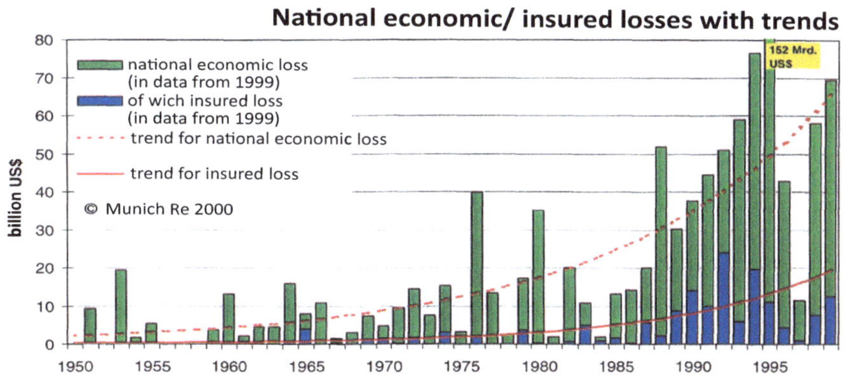

Major natural disasters 1950–1999
decade-by-decade comparison

	decade 1950–1959	decade 1960–1969	decade 1970–1979	decade 1980–1989	decade 1990–1999	factor 90s : 50s	factor 90s : 60s
number	20	27	47	63	86	4,3	3,2
national economic loss	39,6	71,1	127,8	198,6	607,0	15,3	8,5
insured loss	0	6,8	11,7	24,7	109,1	–	16,0

loss in billions of US$ (in data from 1999)

NatCat *SERVICE*
© Munich Re, REF/Geo – January 2000

The figures (from the year 2000!) speak clearly:

- Since the 1970s, national economic and insured loss has increased significantly; loss trends have distinctly and consistently been pointing upwards since this point in time.
- The decade-by-decade comparison demonstrates that the number of natural disasters has more than tripled since the 1960s compared to the 1990s (factor 3.2), while national economic loss has risen more than eightfold and insured loss even by 16-fold.

In addition, the Munich Re records in its publication from the year 2000[36]: "*One look at the development of temperatures over the last 1000 years (Northern hemisphere) clearly shows: since the mid-19th century, a serious increase in temperature has been taking place on our planet. Unless drastic reduction measures for the most important greenhouse gases are taken, the global average temperature could rise by up to 4 degrees in the next 100 years—a value that would have dramatic effects (translation, original quote in German).*" Furthermore, the Munich Re assumed even at that time that storms would increase outside of the tropics as well, that tropical storms (hurricanes, typhoons, cyclones) would increase in intensity and would also occur outside of their traditional regions, that winter gales would become stronger and that there would be an increase in heavy rains and flash floods[37]. Based on this, it concluded[38]: *"It must be feared that new extremes for a multitude of insurance-related parameters are going to appear in almost all regions of the Earth in the course of climate change. This will lead to natural disasters of unprecedented strength and frequency (translation, original quote in German)."* In conclusion, Munich Re states[39]: *"As much as one can still continue to argue about the future development of climate change and also, in particular, its effects, the signs for a future aggravation of the risk situation are irrefutable (translation, original quote in German)."*

The "Kantonale Gebäudeversicherungen" (Cantonal building insurances) in Switzerland (hereafter referred to as "KGV", see p. 46 et seq. below), which insure about 85% of Swiss buildings and about 82% of the insured building capital against damage caused by natural hazards[40], issued the following statement as long ago as 1995[41]: *"Elemental events are increasing in frequency the world over, both in the number as well as in the extent of insured loss. This phenomenon can be observed in Switzerland as well. For example, in the sector of the cantonal building insurances, the average loss as the result of damage caused by natural hazards (10-year average) on buildings has increased by 50% within the last 10 years. The corresponding numbers are even more impressive for single events (translation, original quote in German)."* The KGV reacted and put the Interkantonale Risikogemeinschaft Elementar (hereafter referred to as "IRG") into effect as long ago as 1996. This results in the solidary defrayment of losses in the event of natural disasters or extraordinarily high losses (for more detail, see p. 52 et seq. below). To begin with, an additional capacity of CHF 500 million per year was created through the IRG for claims pay-

---

36 Munich Re, topics 2000, p. 105.

37 Munich Re, topics 2000, p. 107 et seq.

38 Munich Re, topics 2000, p. 111 et seq.

39 Munich Re, topics 2000, p. 112.

40 Annual report Vereinigung Kantonaler Feuerversicherungen (hereafter referred to as "VKF") 2007, p. 2; annual report VKF 2006, p. 7. The VKF is the association of the Kantonale Gebäudeversicherungen.

41 Annual report Interkantonaler Rückversicherungsverband hereafter referred to as "IRV") 1995, p. 5.

ments. From 1999 on, the capacity was increased to CHF 750 million, absolutely on time as it was to turn out, for the KGV were struck by the highest losses caused by natural hazards in 1999 ever experienced in their previous history, about CHF 957 million (the winter storm "Lothar" concluded the year on 26 December 1999 which alone cost CHF 450 million in damage to property)[42].

Hence, climate change is no surprise at all for insurance institutions providing coverage against natural hazards—and especially in Switzerland. On the contrary, they have long adapted to the fact that climate change which has, and is going to have, negative effects, is actually taking place.

B. The effects on insurance against natural hazards in detail

a) Current loss trends

The current trends confirm the findings of the insurances against natural hazards in the late 1990s and are continuing in this framework. The number of natural disasters *in all* continues to rise[43]:

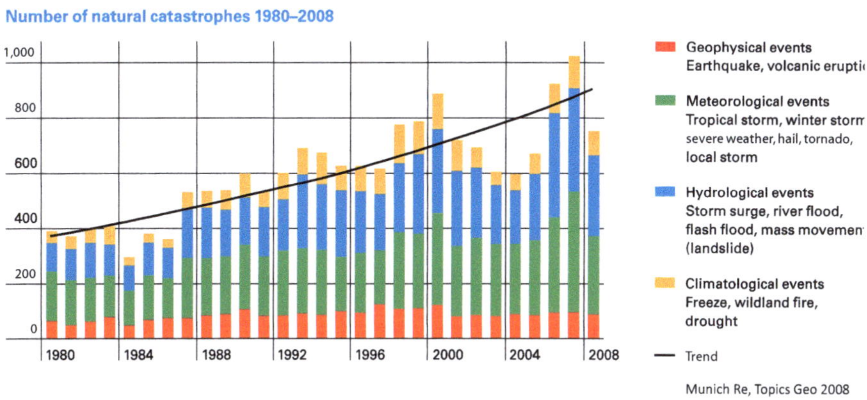

Even the trend regarding the number of great natural catastrphes[44] as well as the trend regarding overall losses and insured losses has steadily been pointing upward since the end of the 1990s[45]:

42 Annual reports IRV 1995, p. 4 et seq., 1998 p. 4 and 1999 p. 4 et seq.; annual reports VKF 1998 p. 29 and 1999 p. 7 et seq. Cf. also WANNER, p. 91 et seq.

43 Source: Munich Re, Topics Geo, Natural catastrophes 2008—Analyses, assessments, positions, Munich 2009, p. 35 (cited Munich Re, topics 2008).

44 A natural disaster is described as "great" if, following the definition by the United Nations, it significantly exceeds the affected region's self-help capacity and if national or international aid is required (Munich Re, topics 2008, p. 38).

45 Source: Munich Re, topics 2008, p. 39.

Number of great natural catastrophes, 1950–2008
The chart shows for each year the number of great natural catastrophes (Category 6), divided up by type of event.

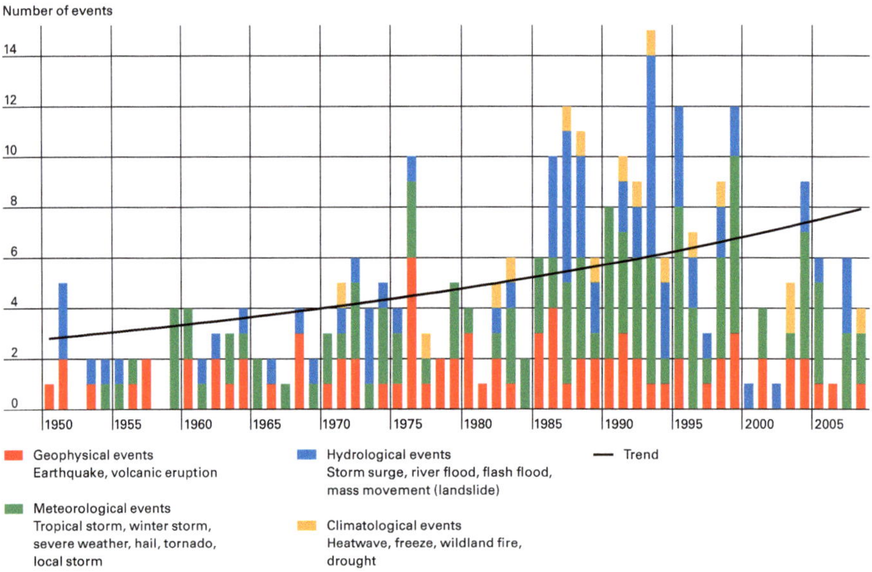

Overall losses and insured losses from great natural catastrophes, 1950–2008
The chart presents the overall losses and insured losses – adjusted to present values. The trend curves verify the increase in Category 6 catastrophe losses since 1950.

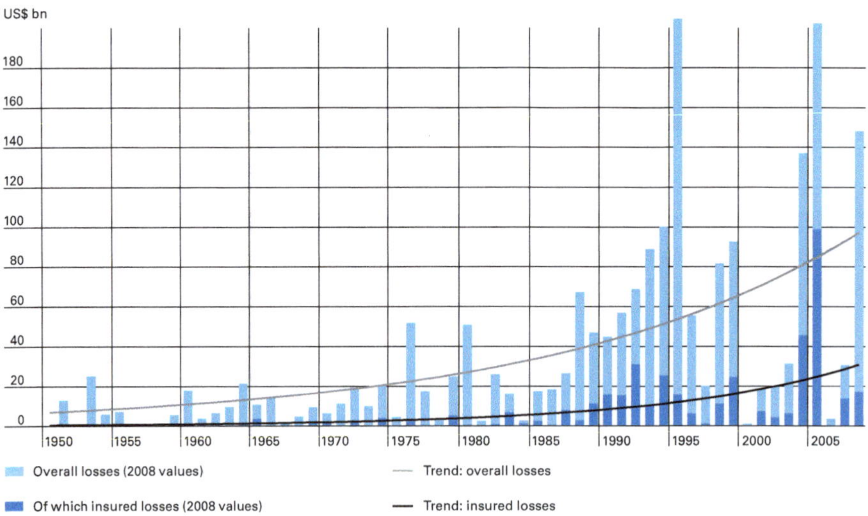

These trends and their confirmation are manifesting and confirmed in Switzerland as well. The following figure by the KGV shows that losses to buildings caused by natural hazards have been steadily increasing since the 1970s[46]:

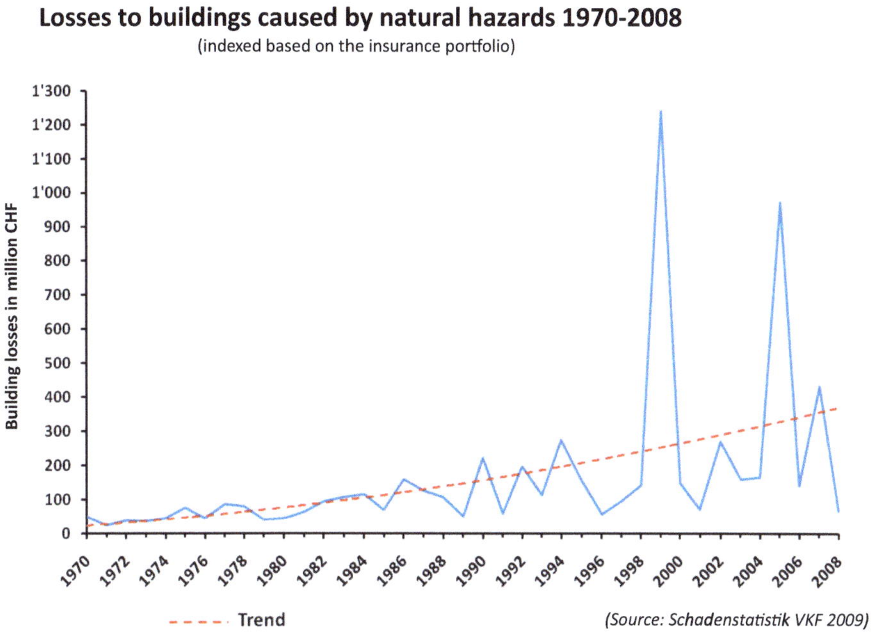

When focusing on individual natural hazards, Swiss Re assumes an increase of 16–68% for the period from 1975–2085 (compared to 1961–1990) based on a study on storm damage in Europe, without taking inflation into account. According to Swiss Re, the change affects both the intensity as well as the frequency of storms. The conclusion drawn by Swiss Re is unambiguous: "*Climate change is going to result in more frequent and more severe winter storms in Europe in the long term and thus in greater losses as well.*"[47]

b) The change in actuarial benchmark figures

Within the scope of the report "Klimaänderung und die Schweiz 2050" (climate change and Switzerland 2050), the insurance sector was examined in more detail. In addition, insurance companies that offer coverage against natural hazards have dealt with the change in benchmark figures. The following changes are to be expected:

- *Loss potentials* are increasing considerably globally and in Switzerland. Insurances against natural hazards must prepare for this. Taking recourse to historical

46 Source: Schadenstatistik VKF, Bern, 2009.
47 Swiss Re, Storm losses, p. 2.

experience figures and statistics will no longer help; instead, future loss potentials must project the forecast climate change with its ramifications. The primary concern is to estimate the worst case annual loss potential—*Possible Maximum Loss (PML)*. The PML must be revised upwards considerably, in Switzerland primarily due to the increased loss potential of the danger of storm, flooding and hail[48]. The PML benchmark figure in turn has a bearing on necessary capitalisation, on premiums, and on the insurance's reinsurance needs, among other things.

- The *Recurrence periods* are getting shorter. Thus, it is assumed, for example, that the recurrence period of a 5-day precipitation amount that leads to flooding in Switzerland will be cut in half, and that a 100-year event will become a 50–100-year event (the corresponding flooding can therefore be expected as often as every 50 years, instead of only every 100 years). This continues on downwards, i.e., a 20-year event becomes a 10–20-year event[49]. The change of this benchmark figure must necessarily also lead to an adaption of risk assessment and risk management.
- Weather events and associated losses that have occurred relatively rarely so far are going to occur so frequently in future that the issue is raised as to whether they are still accidental actuarial loss events, or whether one will have to refer to them instead as regular and thus foreseeable events. Insuring foreseeable events, however, contradicts the insurance principle according to which only accidental, unforeseeable events are covered[50]. Discussions on whether certain events will continue to be insured (e.g. regularly occurring flooding events) are conceivable. It is here in particular where loss prevention comes to the fore as a remedial measure.
- The significance of the *prevention of losses caused by natural hazards* increases significantly. The prevention of losses caused by natural hazards cannot prevent climate change and natural disasters. It can, however, help to limit the negative consequences, the losses[51]. In view of the future, extremely high loss potentials, loss prevention will become a *necessity* in order to avoid the discontinuation of insurance protection due to uninsurability. The extent of losses must be limited from the start. To this effect, a wide variety of means are available as regards building damage, e.g. property protection by means of constructional measures, the use of suitable building materials, the creation and implementation of suit-

48 Report "Climate change and Switzerland 2050", p. 142 et seq; annual report IRV 1995, p. 5 Swiss Re, Storm losses, p. 3. The KGV have examined their PML in 2003 and have concluded that the significantly increased PML of CHF 1,300–1,400 million is sufficiently covered due to the additional IRG capacity of CHF 750 million (cf. annual report IRV 2003, p. 6 et seq.).

49 Report "Climate change and Switzerland 2050", p. 144; Swiss Re, Storm losses, p. 3.

50 Report "Climate change and Switzerland 2050", p. 146 et seq.

51 Report "Climate change and Switzerland 2050", p. 147 et seq.; Swiss Re, Storm losses, p. 2 et seq.

able construction standards and hazard mapping. All of this can be directly or indirectly promoted by insurance against natural hazards (cf. p. 56 et seq. below).

C. The increasing significance of insurance against natural hazards

Natural catastrophes in recent years have drastically demonstrated the huge losses effected by such events. Hurricane Katrina (August 2005) has moreover illustrated the vulnerability even of western industrial nations. The consequences of Katrina have not been overcome as of today, not least due to the lack of, or completely insufficient, insurance against natural hazards (see p. 75 et seq. below).

As the result of the considerable increase in extreme weather events, natural disasters and losses caused by natural hazards, in other words, the increase in the danger of losses caused by natural hazards overall, the necessity to safeguard against the consequences is increasing. A loss event can take on an existence-threatening scale faster than before. Residential buildings just like factories, commercial properties or hospitals can be completely destroyed or severely damaged, making complete reconstruction necessary. The financial means to this end are likely often not available so that insurance provision is required. In any case, the scarcity or the modest extent of any extreme weather events that may occur can no longer be speculated for. It is precisely the accumulation of floods, hail events and storm losses in recent years in Switzerland that has shown that such events are by no means rare any longer and that coverage is therefore a necessary standard, or should therefore become a necessary standard where this is not yet the case.

Based on the huge losses and loss potentials as the result of natural disasters, the relevance of insurance against natural hazards will subsequently be beyond the scope of individual coverage and attain *national economic importance.* If, following relatively extensive devastation, businesses, factories and residential buildings are not rebuilt relatively expeditiously, or if there is even uncertainty as to whether it will be possible to rebuild, this will lead to a collapse of the national economy affected. The inhabitants will move away for lack of means to earn a living or stay behind without any prospects in makeshift accommodations. The case of Katrina may serve as an illustrative example here (see p. 75 et seq. below).

The significance of insurance against natural hazards is therefore going to increase globally, including in Switzerland, due to climate change.

2. Future demands on protection against natural hazards and insurance of buildings against natural hazards

The challenge of climate change leads to a change in the demands on protection against natural hazards and insurance protection against losses caused by natural hazards. In future, far more people and their property are going to be affected by natural hazards, and to a more severe extent. This calls for a strengthened "defence strategy". In order to meet the challenge of climate change, special requirements, including with respect to building protection, must be met:

- Protection should be granted *spatially inclusively* and *comprehensively* since all areas are affected by the various natural hazards (flooding, storm, hail etc.) and this alone will ensure solidarity among all potentially aggrieved parties (see below on solidarity as a necessary element).
- Protection should be *available* to *all* building owners, i.e. all must have the opportunity of acquiring protection, and it must be *affordable* for *all.* Otherwise, rebuilding after a loss event would be compromised.
- Protection should ensure the *rebuilding* of destroyed or damaged buildings *within a useful period of time.* This is of eminent importance for overcoming a natural disaster, including from a national economic point of view (see also below). Natural disasters cause production processes to be interrupted, and consumption declines. Studies have shown that when disasters occur in countries that do not have obligatory insurance, the decline in GDP is greater than if obligatory insurance existed[52].
- Protection should be ensured by means of *insurance against natural hazards.* State aid money or private donations, however, are poor alternatives. Swiss Re has shown convincingly why that is the case[53]. Rebuilding infrastructure and buildings in a timely manner is decisive in overcoming a natural disaster. If there is no right to claim insurance benefits, the aggrieved parties depend on benefactors' goodwill and on political decision-making processes; they become petitioners. If a sufficient sum of financial means is gathered at all, their distribution is not yet regulated, a task that requires time-consuming discussions. In addition, a well-practised claims settlement organisation is lacking, and all this at a time when the catastrophe has already occurred and everyone is waiting for help. The consequence is that the offices involved are completely overwhelmed and frictional losses will be considerable. Rebuilding will be delayed or does not take place. By contrast, an insurance solution allows the distribution of the claims burden to be regulated beforehand, bindingly and quietly, without the pressure exerted by the actual disaster. The aggrieved parties are legally entitled to payments, and discussions on "whether" and "how much" are unnecessary. The financial means will flow and steps can be taken expeditiously towards rebuilding. A *well-practised claims settlement organisation* can record and remedy the losses in a technically irreproachable manner. Moreover, it is significant for the

---

52 Alternative financing and insurance solutions, p. 12 et seq. For this point, the USA were incorporated in this analysis as well.

53 Swiss Re, Floods, p. 2 et seq.

state not to have to use taxpayer money and personnel resources that are urgently needed elsewhere (transportation infrastructure, medical help, etc.) for damage to buildings, thanks to the insurance solution. In addition, it is not necessary for the entire population to pay for damage to property by means of general taxpayer money, although not everyone is a property owner; loss coverage is financed by the affected persons only, namely the community of insured and aggrieved parties (payers of premiums) as well as the insurance, and this is significantly fairer.

- Insurance against natural hazards should therefore provide spatially inclusive and comprehensive insurance protection affordable for all.
- To avoid dependence on donations and state financial aid, at least compulsory insurance (obligatory insurance) is required[54]. Two of the greatest natural disasters in recent years have made the disadvantages of the absence of compulsory insurance perfectly clear (Elbe river deluge/floods in summer 2002 in Germany/Austria/Czech Republic; Hurricane Katrina in August 2005, USA). It turned out that, if compulsory insurance is lacking, numerous buildings will remain uninsured and the aggrieved parties become dependent on donations and state financial aid in a loss event (on the problem of the charity hazard see p. 32 below). In this regard, Swiss Re draws the following conclusion, based on an analysis of the extensive flooding in summer 2002 in Germany, Austria and Czech Republic: "*Since there is virtually nobody who is not exposed to any risk of flooding, mandatory insurance*[55] *would have some advantages, particularly where, as is the current practice in Germany for example, the additional tax burden is also placed on the entire population. Those who have obtained voluntary insurance against the risk of flooding, or who have given donations for flood victims, even have to bear a double burden (translation, original quote in German).*"[56] Kunreuther/Michel-Kerjan record, after an in-depth analysis of the consequences of Katrina (see also p. 75 below): "*Mandatory coverage would address many difficult problems. Since all residents would be financially protected, it would reduce the demand for government aid that is sure to arise after another disaster.*"[57]
- Accessibility of insurance protection for all requires that premiums remain affordable throughout, even for bad risks. Premiums should thus be relatively low or affordable even for bad risks.

---

54 The terms "Versicherungspflicht" (compulsory insurance) and "Versicherungsobligatorium" (obligatory insurance) are used as synonyms here and stand for the legal obligation to obtain insurance protection.

55 When referring to "Pflichtversicherung" (statutory insurance), Swiss Re means obligatory insurance in this case. Especially in Germany, however, "Pflichtversicherung" means a monopoly insurance where everybody must be insured due to obligatory insurance.

56 Swiss Re, Floods, p. 4.

57 Kunreuther/Michel-Kerjan in: Richardson/Gordon/Moore II, p. 29.

- To be able to overcome high losses while at the same time keeping premiums at a level that is affordable by all, the *risk collective* must be as *big* as possible. Studies by Swiss Re have shown that for an extensive risk collective, the standard premium is lower by a factor of 20 than when the risk collective is limited to bad risks ("*The bigger the risk collective, the smaller the burden for the individual (translation, original quote in German)")*[58]. Insurance must therefore operate according to the *principle of solidarity*, which allows *balancing of good* and *bad risks*[59]. Otherwise, the insurance is not able, for lack of a sufficiently big risk collective, to defray the losses, in particular if the insurance community consists primarily of bad risks (on the problem of negative selection or anti-selection see p. 38 below). The possibility of balancing with good risks is lacking, the formation of cluster risks is favoured. Solidarity must thus be as extensive as possible. With respect to the question of whether a monopoly of insurance against natural hazards is permissible to ensure solidarity, we refer to the remarks on pp. 83 et seq. and 104 et seq. below.
- In order to guarantee such solidary insurance against natural hazards in the long run, and also so as not to push the solidarity too far, the *prevention against loss caused by natural hazards* in addition receives an extremely important role (on the problem of moral hazard, see p. 32 below). Due to the huge extent of loss and loss potentials, direct action is required with regard to the extent of loss, in addition to solidary compensation. Loss prevention is designed to achieve a situation whereby losses do not even arise, or arise only to a much lesser extent. This allows a low premium level, or solidary insurance, to be maintained, thus benefiting the building owners (and hence the tenants as well) directly[60]. Prevention should be directly implemented and promoted by insurance against natural hazards, be it by incorporating incentives into the insurance or by the direct advancement of constructional and spatial planning measures.

58 Swiss Re, Floods, p. 4.
59 Likewise, Swiss Re, Floods, p. 2 et seq. Cf. also Swiss Re, Climate change, p. 23.
60 Cf. p. 24 above, p. 56 et seq. below as well as Swiss Re, Climate change, p. 23.

Chapter 2: Insurance systems today

I. Types of insurance systems

1. General information

The answer to the threat of natural disasters on building values will turn out differently depending on the country. Several types can be distinguished based on the relevant characteristics of insurance systems.

Subsequently, the specific insurance systems of a variety of countries will be introduced and allocated to the corresponding types. This will allow a transparent discussion to be held on which types have what advantages and disadvantages.

2. Type 1: Statutory insurance

Statutory insurance refers to the obligation of the building owner to take out insurance against natural hazards, namely *from a specific insurance organisation.* This insurance organisation *has a legal monopoly.* This means two things. First, the building owner must take out insurance from this particular insurance organisation and cannot take out insurance elsewhere. Second, the monopoly insurer cannot reject the potential policyholder, i.e., the insurer is forced to accept the risk. This is therefore a case of compulsory insurance (obligatory insurance) that is implemented by monopoly insurance. Since all building owners are included, the maximum possible risk collective is created and thus the *maximum possible solidarity* among the insured.

3. Type 2: Compulsory insurance (obligatory insurance)

Compulsory insurance refers to the obligation to take out insurance against natural hazards. This also subsumes the situation where the obligation is not created until another insurance agreement (e.g. fire insurance as a basic agreement) is entered into. Strictly speaking, an compulsory insurance no longer exists if this basic agreement was entered into voluntarily, for in this case it can be circumvented by waiving the basic agreement; insurance protection is absent. If this set-up is nonetheless subsumed under the heading of compulsory insurance, it is because, at least in Europe, entering into a basic agreement (as a rule, fire insurance in the form of a household or BII insurance[61]) is so widespread that compulsory insurance is approximately achieved. The term "obligatory insurance" is used synonymously with compulsory insurance, even though the two terms are understood to mean something different in some cases in this respect[62]. According to Swiss legal understanding, obligatory

---

61 Business interruption insurance.

62 Cf. for example the classification in Schwarze/Wagner, in: Märkte und Politik, p. 177 et seq., who describe obligatory insurance as merely the above-described obligation of insurance against natural

insurance describes the obligation to obtain insurance protection independent of a basic agreement.

The main difference to type 1 consists of the fact that compulsory insurance is not effected by a monopoly insurance but by different insurances. Thereby, several smaller risk collectives are created. Frequently, insurance against losses caused by natural hazards cannot be formulated freely, particularly if it is coupled with a basic agreement; instead the scope of coverage and the premium are provided for by law.

4. Type 3: State aid and optional insurance according to market supply

There is no compulsory insurance for this type, not even based on a basic agreement (and thus, there is no basis for statutory insurance either). Whether insurance protection exists is instead at the discretion of the building owner. He decides whether he wants to take precautions by taking out insurance or not, *provided that insurance is offered on the market at all.* Since there is no compulsory insurance and much less so any statutory insurance, the insurance industry is also not obligated to offer insurance protection. Whether insurance protection exists at all is accordingly left to the market forces. In a loss event, and all the more in a disaster situation, state aid is of particularly eminent importance. The claims burden is borne primarily by the state by means of state aid; the aggrieved party is not legally entitled to it.

At most, small risk collectives arise in the absence of statutory insurance and compulsory insurance, if they arise at all.

II. Germany

1. State aid and insurance option according to market supply

A. Market supply in residential buildings as well as supplementary insurance according to the ZÜRS

There is no compulsory insurance in Germany and thus also no statutory or monopoly insurance for natural hazards. Even if a household or BII insurance were taken out, there is no obligation to extend coverage to elementary hazards. The owner of a building decides on a voluntary basis whether he wants to take precautions against natural hazards. Whether it will even be possible at all to take out insurance (a suf-

hazards to be linked to a basic agreement; likewise, Alternative Finanzierungs- und Versicherungslösungen, p. 10. It is, however, stated in both cases that this results in compulsory insurance.

ficiently large offering on the part of the insurers is required as well), is left to the market forces alone. Specifically, the following market supply is available[63]:

- Storm and hail: these risks are regularly included in residential building (household) and BII insurance. Thus, insurance protection exists provided that the appropriate insurance has been taken out.
- Flood/deluge, heavy rains, landslides, snow pressure, avalanches, earthquakes: insurance protection exists only if a separate insurance against natural hazards has been taken out as a supplementary insurance. There is no obligation to grant this protection automatically in terms of a linkage if, for example, a residential building insurance has been taken out. Even if the owner of the building is willing to take out such a supplementary insurance, he will receive insurance protection only if the risk is classified as insurable according to the ZÜRS zoning system, that is, if insurance coverage is even offered on the market in the first place, and this is not a given. ZÜRS was created by the Gesamtverband der Deutschen Versicherungswirtschaft (GDV).
- ZÜRS: the zoning system for flooding, backwater and heavy rain subdivides all of Germany into four hazard classes ("HC") according to recurrence periods (1: more than 200 years; 2: more than 50–200 years; 3: 10–50 years; 4: less than 10 years). Thus, HC 4 means that a loss event, in particular flood damage, takes place more frequently than once in 10 years. Buildings in HC 4 are *not insurable.* The insurance industry simply does not offer any insurance protection, there is no market supply. This means that in a loss event the affected building owners must hope for state aid exclusively.

As shown above already, climate change results in an increase in the loss potential. Recurrence periods are shortened; areas that have thus far experienced little damage are increasingly affected by damage caused by natural hazards. This is bound to affect the ZÜRS, and it remains to be seen whether the HC 4 is extended to include other areas and buildings that will thus become *uninsurable*[64]

---

63 Brochures by the Gesamtverband der Deutschen Versicherungswirtschaft (GDV) "Land unter—Schutz vor Überschwemmung und Hochwasser" and "Stürmische Zeiten—Schäden vorbeugen und richtig versichern", version as published on and downloaded from www.gdv.de/Themen on 25 August 2009. In addition, detailed information is available from the website of the Bavarian state government, www.elementar-versichern.bayern.de/naturgefahren.html (information as per 25 August 2009). In the brochures listed above, the insurance system is described as well. On the insurance system cf. also WEISS, Handbuch Naturkatastrophenrecht, p. 386; SCHWARZE/WAGNER, in: Märkte und Politik, p. 186 et seq.; CONSORCIO, Diversity of Systems, p. 71 et seq.; VW 2000, Hochwasser, p. 275.

64 Cf. also SCHWARZE/WAGNER, in: Märkte und Politik, p. 185 et seq.

B. Ad hoc state aid

Since there is no compulsory insurance at all, it is left to the owner of a building whether he wants to obtain insurance protection or not. If he has not taken out insurance, then he is dependent on state aid in Germany in a loss event. The same holds true if, although the owner of the building would like to take out insurance coverage, none is available on the market because insurances are classifying the risk as being uninsurable. The aggrieved party is thus dependent on payments that it is not legally entitled to. With respect to the remaining adverse consequences of state aid, we refer to the discussion on p. 26 above and p. 79 et seq. below. State aid is not institutionalised but is delivered on an ad hoc basis through government decisions; this, in addition, magnifies the uncertainty.

In addition, the expectation of receiving state aid leads to negative incentives for the owner of a building who will be tempted to rely on state aid instead of purchasing insurance protection (charity hazard). Why should money be spent on insurance premiums if the government is going to pay the loss anyway? Furthermore, state aid also undermines the motivation of the owner of the building to take loss prevention measures (e.g. constructional adjustments) (moral hazard). Why should the owner of a building invest in prevention measures if the government will come to the rescue anyhow and absorb the loss? The building owner's own provisions, which are an important element in loss reduction and in reducing the need for state aid, is itself undermined by state aid.[65]

The Federal states have recognised this problem. In the Federal state of Bavaria, for example, state aid is principally paid out only to those aggrieved parties whose houses were not insurable due to a lack of market supply. This is designed to counteract the charity and moral hazard and put pressure on the owner of the building to take out insurance. Thus, in any case, aggrieved parties remain dependent, however, on state aid where there is no market supply. Furthermore, aggrieved parties that have not obtained insurance protection even though insurance was offered on the market, will demand state aid nonetheless; ultimately, it is politics that decides on whether these aggrieved parties are awarded state aid even though they neglected to take their own precautions.

The German insurance system corresponds to type 3 of the types of insurance systems presented.

[65] Cf. Alternative Finanzierungs- und Versicherungslösungen, p. 10 et seq.

2. Comparison with the situation of the German statutory insurances until 1994 and the Swiss statutory insurances

A. Comparison with the situation of the German statutory insurances until 1994

In Germany, the fire risk was covered by the twelve state monopoly insurances of the various federal states until 1 July 1994 within the scope of statutory insurances. Two of these monopoly insurances, namely the Badische Gebäudeversicherung and the Württembergische Gebäudebrandversicherung, extensively covered natural hazards as well. In the course of implementing the Third EU non-life insurance Directive[66], Germany abolished these insurance monopolies as of 1 July 1994[67]. With respect to the question of whether the revocation of these monopolies was even legally required, we refer to p. 104 et seq. below.

In order to remain competitive, the former monopoly insurers were unable to maintain their extensive coverage against natural hazards. The consequences of the revocation of the insurance monopolies or statutory insurance were and are negative for building owners[68]:

- Administrative costs for building insurers rose between 60% and 100%.
- Premium levels rose by 40% to 80%.
- Expenditures for prevention were cut by about 50%.
- Bad risks were barred from insurance.
- Excesses were increased.

B. Comparison with Swiss statutory insurances

In August 2005, considerable flood damage occurred as a result of heavy rain in the tri-border region of Switzerland-Austria-Germany. The Canton of Grisons, in particular, in Switzerland, Vorarlberg and Tyrol in Austria, and Upper Bavaria in Germany were affected. Since the regions were affected by the same event, and since they have a similar topography and were affected to the same extent, a comparison can be made as to how the respective insurance systems overcome this natural disaster. The special situation allowed an empirical examination of the efficiency of the different insurance systems. In the Canton of Grisons, building values are insured via a statutory insurance, the state building insurance organisation of the Canton of Grisons, which has a monopoly.

---

66 Directive 92/49, see footnote 297 below.

67 Von Ungern-Sternberg, p. 144 et seq., 149 et seq.; Quinto, p. 29.

68 Von Ungern-Sternberg, p. 157 et seq.; VKF annual report 2007, p. 21.

Specifically, loss coverage and insurance density as well as the promptness of claims settlement were compared[69].

a) Loss coverage and insurance density

- Canton of Grisons: Since all buildings were included in the statutory insurance, the insurance density was 100%, and 100% of the damage could be covered.
- Upper Bavaria: Insurance density was only about 3.3%, i.e. only about 3.3% of buildings had the appropriate insurance protection. Furthermore, Bavaria was relatively consistent with respect to paying state aid. Owners of buildings whose buildings were insurable and who did not take out insurance coverage, were not paid any state aid. 18.3% of the damage was covered by private insurance against natural hazards, 22% was covered by state aid payments. This left behind a "hole" of 60% of damage.

b) Promptness of claim settlement

Specifically, the interval between the time the loss was reported and the time the first compensation was paid were compared:

- Canton of Grisons: The first payment was made within 14 days in 50% of cases, within 21 days in 90% of cases (average).
- Upper Bavaria: 38 days passed in 50% of cases, 135 days in 90% of cases (average).

Accordingly, the insurance system in Upper Bavaria required 6.4 times as much time compared to the Swiss system.

c) Conclusion

The comparison shows that the statutory insurance system is far superior in overcoming the natural disaster compared to the market solution with state aid. Statutory insurance naturally results in a much greater insurance density and thus in far greater loss coverage. No damage is left undone, and the owner of a building is left neither to his own devices nor to state aid. In addition, claims are settled much faster. This is not surprising, for both the claims as well as the organisation of claims settlement are prespecified.

---

[69] Alternative Finanzierungs- und Versicherungslösungen, p. 14 et seq., p. 15 et seq.

III. France

1. Compulsory insurance in the sense of obligatory extension of coverage

The French system of overcoming natural disasters is a combination of government intervention for direct insurance and government reinsurance. The owner of a building is not obligated to take out insurance against natural hazards directly. If the owner of a building does, however, take out any household or BII insurance (e.g. building and household insurance against fire and burglary), coverage must compulsorily be extended to include the following hazards (statutory extension of coverage)[70]:

- Storm, hurricane and cyclone. Hail, snow pressure and frost, on the other hand, should already be included in the standard coverage of the property insurance;
- Natural disaster hazards, namely floods, mudslide, landslides, earthquakes, submergence of ground (as the result of desiccation), storm surges (ocean), water/mud/lava flows, moving snow and ice masses. For the French overseas territories, storms that reach average speeds of at least 145 km/h (for ten minutes) or gusts of at least 215 km/h are considered natural disaster risks as well. The list is not conclusive. It is designed to insure the typical natural disaster risks commensurate with French requirements that are considered uninsurable on the common market.

The corresponding statutory, formulated coverage clauses must be incorporated into the basic agreement[71].

There is *no* obligation to conclude a basic agreement, that is to say, property insurance or BII insurance. Thus, no protection exists if *no* such basic agreement has been concluded. Hence, the policy can be circumvented (by the owner of the building). Since the basic agreements mentioned above are, however, very common, the system will nevertheless be referred to here as compulsory insurance (see p. 29 et seq. above). Furthermore, the private direct insurer is under an obligation to contract regarding statutory extension of coverage when a basic agreement is entered into. If the private direct insurer refuses to follow the statutory extension of coverage, for example by not incorporating the coverage clauses, the rejected insured party can involve the Bureau Central de Tarification ("BCT"). The BCT is a state regulatory authority that enforces compulsory insurance in a wide variety of sectors (e.g. also

---

70 Art. 1 Loi no. 82-600 due 13 juillet 1982 relative à l'indemnisation des victimes de catastrophes naturelles; Art. L 122-7 Code des assurances; Art. L. 125-1 et seq. Code des assurances; Art. 13 loi d'orientation pour l'outre-mer du 13 décembre 2000; CCR, Catastrophes Naturelles, p. 5 et seq., 18, 21.

71 Art. 2 Loi no. 82-600; Art. L125-2 Code des assurances; Art. A125-1, Art. Annexe I art. A125-1 and Art. Annexe II art. A125-1 Code des assurances.

with respect to the mandatory motor vehicle indemnity insurance). The BCT examines the insured party's complaint and compels the specific direct insurer, under penalty of having his licence withdrawn, to grant the extension of coverage[72].

In addition to concluding a basic agreement, another requirement must be met so that compensation can be paid *due to the occurrence of natural disaster hazards.* The occurrence of a natural disaster must *be declared by an interdepartmental decree.* The mayors of the affected municipalities put together a report for the attention of the department prefect. The latter sends a corresponding, comprehensive dossier to an interdepartmental commission with a motion to declare a natural disaster. In its decree, which is published in the Journal Officiel de la Republique Française, the commission specifies which regions are affected, the time period of the natural disaster, as well as the natural hazards and types of losses that have occurred[73].

Furthermore, the following should be stated for natural disaster coverage[74]:

- The insured property and the insurance sum depend on the basic agreement. Thus, the building must be insured against fire, for example, in the basic agreement, so as to be insured against natural disasters as well. Compensation is paid according to the value or according to the compensation principle in the basic agreement. If replacement value coverage is provided for in the basic agreement, this applies equally therefore even if the building is destroyed by a mudflow.
- The premium is set by decree as a percentage of the premium of the basic agreement and is currently 12% for buildings. It is a *state-regulated standard premium,* i.e., no gradation takes place based on the building's risk exposure. This facilitates and strengthens *solidarity* among insured parties.
- The excess is €380.

Even in France, the necessity of *preventing losses caused by natural hazards* has been recognised against the background of increasing natural disasters. In order to finance the corresponding measures, the Fonds de Prévention des Risques Naturels Majeur ("FPRNM") is increased continuously. Part of the premium for coverage of natural disasters is transferred to the fund for the purpose of monetary injection, namely currently 8% of the premium (thus, the contribution is included in the premium and, accordingly, fewer funds are available for covering losses); the

[72] Art. L 125-6 and R 250-1 to R 250-6 as well as A 250-1 and A 250-2 Code des assurances; CCR; Catastrophes Naturelles, p. 8 et seq.

[73] Art. 1 Loi no. 82-600 du 13 juillet 1982 relative à l'indemnisation des victimes de catastrophes naturelles; Art L 125-1 al. 4 Code des assurances; CCR, Catastrophes Naturelles, p. 5, 11, 18; Consorcio, Diversity of Systems, p. 62 et seq.

[74] Art. L 125-2 al. 3, Art. Annexe I and Art. Annexe II art. A125-1 Code des assurances; CCR, Catastrophes Naturelles, p. 6 et seq.; Consorcio, Diversity of Systems, p. 64.

contribution can be increased to 12%[75]. In addition, the same is fixed for the excess and pressure is exerted in this way on the municipalities and on owners of buildings. The municipalities must develop a Plan de Prévention des Risques naturels prévisible ("PPR") that contains and implements specific loss prevention measures with respect to foreseeable natural hazards. If a municipality does not have a PPR and if a natural disaster occurs, it is examined how often, prior to the corresponding interdepartmental decree, other decrees had been issued for this municipality already which referred to the same (foreseeable) natural disaster (e.g. flooding). The excess is then multiplied based on a scale. If it turns out, for example, that, after renewed flooding, a natural disaster due to flooding has already been recorded in three decrees prior to the decree in question, and if nevertheless no PPR is on hand, the excess is tripled; these excesses are mandatory, i.e., they will also be applied if the basic agreement does not contain such a provision, and they cannot be absorbed by separate insurance[76]. In addition, the state re-insurance Caisse Centrale de Réassurance ("CCR") promotes the prevention of loss caused by natural hazards as well.

2. Effective defrayment of loss in the natural disasters sector by the state reinsurance Caisse Centrale de Réassurance (CCR)

For the owner of a building, the natural disaster risk is "insurable" only because it can be widely passed on to the state reinsurance, the CCR, by private direct insurers. Although the CCR has the legal structure of a stock company, 100% of the share capital is, however, owned by the state[77].

Although the CCR does not have a legal monopoly with respect to reinsuring natural disaster risks, i.e. the direct insurers can cover their natural disaster portfolio with any reinsurer, the terms and framework conditions of the CCR are, however, so advantageous that assigning it to a private reinsurer makes no sense. In particular, the CCR can absorb the natural disaster risk without limits and is itself safeguarded by an unlimited state guarantee[78].

Reinsurance by the CCR takes place in the following framework[79]:

---

75 Loi no. 95-101 du 2 février 1995 relative au renforcement de la protection de l'environment; Loi de Finance no 2008-1425 du 27 décembre 2008; Rapport Annuel 2008 of the CCR, p. 23.

76 Art. Annexe I and Annexe II art. A125-1 as well as Art. A125-3 Code des assurances. CCR, Catastrophes Naturelles, p. 7; Consorcio, Diversity of Systems, p. 65.

77 Art. L431-9 Code des assurances, Art. 11 Loi no. 92-665 du 16 juillet 1992; CCR, Catastrophes Naturelles, p. 3; Consorcio, Diversity of Systems, p. 67;

78 CCR, Catastrophes Naturelles, p. 3, 19 et seq.; Art L431-9 Code des assurances; Décret no 82-706 du 10 août 1982; Consorcio, Diversity of Systems, p. 66 et seq.

79 CCR, Catastrophes Naturelles, p. 3, 19 et seq.; Consorcio, Diversity of Systems, p. 67 et seq.

- Quota share reinsurance: the direct insurer can currently transfer exactly 50% of his *entire* natural disaster portfolio to the CCR, while the remaining 50% remain with the direct insurer as a priority (point of attachment) (on the problem of negative selection see below).
- Stop-loss reinsurance: above an agreed annual loss sum (priority/annual aggregate), the loss is absorbed by the CCR (annual excess of loss); there is no limit regarding the defrayment of loss by the CCR.

Even the CCR continually strives to assess the damage-causing event and to limit the potential extent of losses a priori by means of prevention measures. For this purpose, hazard maps are prepared in order to render risk exposure and loss potential more transparent[80]

Finally, it should be noted that the CCR does not only operate the reinsurance of natural disaster risks with a state guarantee, but reinsures other risks as well, without state guarantee. In this respect, the CCR differentiates between "réassurance avec garantie de l'État" and "réassurance de marchés". The latter, however, should not amount to more than one third of the entire premium revenues, while reinsurance of natural disaster risks should continue to represent the focus of the CCR's activity, amounting to two thirds. The CCR at any rate consistently makes sure that its premium revenues from "market reinsurance" do not exceed the rate of one third[81] .

The French insurance system corresponds to type 2 of the types of insurance systems presented.

3. Adaption of the French system due to moral hazard and negative selection

The current framework conditions of the French system presented above are the result, not least, of numerous adaptions made in recent years. These adaptions had to be undertaken because the French system was affected to a considerable extent by the problems of moral hazard and negative selection which threatened to unhinge the entire system. *Negative (risk) selection (also described as "anti-selection" or "adverse selection")* refers to the problem whereby only insured parties with high risks (*bad* risks) take out insurance, which will lead to an accumulation of bad risks with the insurance organisation and thus to financial imbalance. The *good* risks, however, do not take out any insurance at all and/or have alternatives that are not open to bad risks[82].

80 Rapport Annuel 2007 of the CCR, p. 7, 17.

81 Rapport Annuel 2005 of the CCR, p. 23; Rapport Annuel 2007 of the CCR, p. 6 et seq., p. 11 et seq.; Rapport Annuel 2008 of the CCR, p. 11 et seq.

82 Cf. von Fürstenwerth/Weiss, p. 443; Swiss Re, Manual, p. 243.

This development is hardly astonishing in view of the increase in natural disasters and is a quintessential example of what basic requirements an insurance system will have to meet in order to be able to persist in the light of climate change. The problems that have occurred and the adaption measures are briefly presented below:

- *Moral hazard:* due to the relatively comfortable insurance protection and the low excess, there was too little incentive to pursue loss prevention. This resulted in repeated losses taking place in the same regions, although they could have been prevented or at least limited by preventive measures. This is where the ruling effective as of 1 January 2001 regarding the multiplication of excesses in the absence of a PPR (which cannot be bought back) had a counteractive effect (see p. 37 above). If no preventive measures are taken following a loss event, a higher excess will have to be paid. The contribution to increase the FPRNM mentioned above, the funds of which were invested in loss prevention, aims in the same direction. The contribution was increased in three increments from 2% to 8% from 1999–2008 and can now even be increased to as much as 12%.
- *Negative selection for storm risk:* it was realised as long ago as the 1980s that, if anything, only those owners of buildings exposed to a significant storm risk will take out insurance. This resulted in an accumulation of bad risks with insurances. Storm insurance met with little acceptance, both on the part of the owners of buildings as well as on the part of the insurances, a fact that resulted in uncovered losses in loss events. Therefore, statutory extension of coverage to include storm risk was introduced in 1990[83].
- *Negative selection at the disadvantage of the CCR:* originally, direct insurers were able to transfer up to 90% of their natural disaster risks to the CCR. In addition, it was possible to transfer only parts of the portfolio, even risks from individual contracts only. This resulted in direct insurers separating out their bad risks and transferring them to the CCR as a whole. Direct insurers with predominantly good risks, however, did not transfer anything, or only a small quota (the minimum quota was 40%) to the CCR, and even here bad risks were separated out as well. This resulted in a considerable accumulation of bad risks at the CCR. This had the effect of undermining solidarity, if not even making it impossible. Balancing of good and bad risks could not take place. In 1999, the CCR had to draw from the state guarantee in the amount of FF 3 billion. Fluctuation reserves were severely depleted and had to (or have to) be rebuilt. From the year 2000 on therefore, various adjustments were made in order to limit negative selection and to restore the financial equilibrium of the CCR. In particular, only 50% of risks (or even 0%) can be transferred to the CCR now, and these 50% must relate to

83 Loi no 90-509 du 25 juin 1990; Art. L122-7 Code des assurances. CCR, Catastrophes Naturelles, p. 18; Consorcio, Diversity of Systems, p. 65; von Ungern-Sternberg, p. 93 et seq.

the *entire* portfolio. Separating out bad risks to the disadvantage of the CCR is no longer possible. Even so, the CCR still does not have a monopoly, so that a direct insurer who has many good risks is free not to choose the CCR as a reinsurer. The CCR's focus is on the efforts to restore the financial equilibrium of the CCR, and these efforts are continuing to this day[84].

The following conclusions can be drawn from the experiences in France with respect to demands on an insurance system that is designed to withstand climate change:

- The *prevention of negative selection* and the securing of *solidarity* among the insured parties are a *prerequisite* for a properly functioning insurance system. Otherwise, it will quickly *lose its financial equilibrium* and losses can no longer be defrayed due to the lack of sufficient funds.
- The larger the set of risks incorporated into the insurance system, the sooner negative selection is avoided and solidarity can be ensured. Good and bad risks must be included in equal measure since this is the only way that equilibrium between risks can be ensured.
- *Incentives* for *loss prevention* must be created by the insurance system. The system and the insured parties must have an immediate interest in preventing losses. Due to the enormous loss potential of climate change, the direct prevention of losses, in addition to insurance, is imperative. Losses that strain the community of insured parties should not even arise in the first place.

4. Differences to the German and the Swiss as well as the Spanish system

The French system find itself in the middle between the German solution on the one hand and the Swiss or the Spanish solution on the other:

- In Germany, the problem of overcoming losses caused by natural hazards is left to the private insurances market supply and ad hoc state aid. In contrast to France, there is neither compulsory insurance nor a state-owned insurance organisation that covers the losses.
- Switzerland (for by far the largest part of its building stock) and Spain, on the other hand, go even one step further than France. There is not only compulsory insurance but also statutory insurance whereby maximum solidarity among insured parties can be ensured. Furthermore, neither in Switzerland nor in Spain does a natural disaster have to be declared by the state before insurance benefits

84 CCR, Catastrophes Naturelles, p. 14, 19 et seq.; CONSORCIO, Diversity of Systems, p. 67 et seq.; VON UNGERN-STERNBERG, p. 105 et seq., 113 et seq.

can be claimed. Once the natural disaster has occurred, a loss report submitted by the aggrieved party is sufficient to release the insurance claim.

IV. Spain

1. Statutory insurance for extraordinary risks

A. State-owned enterprise Consorcio de Compensación de Seguros as risk carrier

In Spain, extraordinary risks are insured by the Consorcio de Compensación de Seguros (hereafter referred to as "Consorcio"). The Consorcio is an insurance organisation regulated by public law, with its own legal personality and its own assets, which is led by an administrative board. The members of the administrative board are appointed by the Spanish Secretary of Trade and Commerce, and the director of the Spanish insurance supervisory office holds the office of president of the administrative board by operation of law[85]. Although the Consorcio is in principle subject to private insurance supervision and its contracts are subject to the Insurance Contract Act, all this is subject to the provisions of the Estatuto Legal des Consorcio (hereafter referred to as "Estatuto Legal") and further decrees supported by it, as well as other public-law regulations; the Estatuto Legal is the decree governed by public law that constitutes the Consorcio. This shows that all essential points of the insurance relationship, in particular insured risks, premiums and insurance benefits, are regulated by the Estatuto Legal and the regulations that support it, that is, by public law[86] (issues such as the creation of fluctuation reserves, which in substance also come under regulatory law, are regulated by the Estatuto as well[87]). As a result, insurance through the Consorcio thus complies with special, public law.

B. Statutory insurance due to statutory coverage clause and statutory surcharge

The Consorcio insures persons and property located in Spain (in particular buildings in Spain) against *extraordinary risks.* This includes, among other things, the following natural hazards[88]:

---

85 Art. 1, 2 and 4 Estatuto Legal del Consorcio de Compensación de Seguros dated 19 December 1990 [Bylaws of the Consorcio de Compensación de Seguros dated 19 December 1990] according to Ley (law) 21/1990 de 19 de diciembre 1990 with changes according to Ley no 12/2006 de 16 de mayo 2006 and according to Real Decreto Legislativo (royal statutory decree) 7/2004 de 29 de octubre 2004 (cited Estatuto Legal).

86 Art. 6 et seq., 18 Estatuto Legal.

87 Art. 24 Estatuto Legal.

88 Art. 6 Estatuto Legal.

Earthquakes, seaquakes, floods, volcanic eruptions, hurricanes, meteorite impacts. The risk of flooding is by far the most important, averaging over 90% of compensation payments[89].

The Consorcio covers the corresponding personal and property damage as well as the lost profits[90]. It represents a *statutory insurance* and accordingly has a *legal monopoly.* Its statutory or monopoly insurance character is not readily recognisable. According to art. 8 par. 1 a) Estatuto Legal, the Consorcio will pay insurance benefits only "*if the extraordinary risk covered by the Consorcio is not covered by the insurance policy* [of the private insurance]"[91]. *The inclusion of private insurances could convey the impression that the Consorcio grants coverage only subsidiarily to the first one. This impression is wrong, however, for its statutory and monopoly character arises from the following regulations and conditions:*

- Even art. 8 no. 1 a) Estatuto Legal refers to the risk covered by the Consorcio, i.e. coverage by the Consorcio exists in any case; the Consorcio will, however, not make any payments if the private insurance itself has covered the risk as well.
- Coverage by a private insurance, however, makes no sense, as each private property and personal insurance policy (from the lines of business and industries listed in the Estatuto Legal, in particular fire insurance for buildings) must incorporate a coverage clause from the Consorcio as required by law. Coverage by the Consorcio is contained in this literally statutory clause[92]. It is a statutory extension of coverage where the insurance underwriter (Consorcio) is specified as well.
- Each private insurance must raise a premium in the form of a *surcharge* ("recargo") for the Consorcio coverage on each property and personal insurance

---

89 Consorcio, Diversity of Systems, p. 137.

90 Art. 6 Estatuto Legal.

91 "Que el riesgo extraordinario cubierto por el Consorcio no esté amparado por póliza de seguro."

92 Art. 8 no. 3 Estatuto Legal; Art. 4 and 12 Reglamento del Seguro de Riesgos extraordinarios [decree on the insurance of extraordinary risks] (approved by Real Decreto no 300/2004 de 20 de febrero de 2004 with changes according to Real Decreto no 1265/2006 de 8 de noviembre de 2006) (cited Reglamento del Seguro); Art 1.1 and appendix II Resolución de 27 de noviembre de 2006, de la Dirección General de Seguros y Fondos de Pensiones, por la que se aprueban los recargos en favor del Consorcio de Compensación de Seguros en materia de seguro de riesgos extraordinarios a satisfacer obligatoriamente por los asegurados, la cláusula de cobertura a insertar en las pólizas de seguro ordinario y la información a facilitar por las entidades aseguradoras relativa a las pólizas incluidas en el régimen de cobertura de los riesgos extraordinarios [decree by the supervisory authority of insurances and pension funds dated 27 November 2006 to approve the obligatory surcharges to be paid by insured parties for the benefit of the Consorcio de Compensación de Seguros in the sector of insurance of extraordinary risks, the coverage clause to be inserted into the usual insurance policies and the information regarding the policies which are included in the ruling on the coverage of extraordinary risks to be communicated to the insurance companies] (modificada por Resolución de 12 de noviembre de 2008 de la Dirección General de Seguros y Fondos de Pensiones) (cited Resolución 27/11/2006).

policy. Through the private insurance, the surcharge must be collected compulsorily from the policyholder and paid to the Consorcio. It is passed on to the policyholder, i.e., the policyholder pays the surcharge, not the private insurance[93].

- The surcharge must be paid invariably, even if the private insurance also covered the extraordinary risks. According to the express regulation, the surcharge qualifies as tax *("ingresos de derecho público")*[94]. It represents a claim deed under public law and can accordingly be enforced directly[95].
- Since the surcharge must be paid and since coverage by the Consorcio exists in any case, coverage by a private insurance does not make any sense. The latter would then have to pay benefits in a loss event, although the Consorcio would cover the loss as well, for which a surcharge has to be paid in any case. Why should a private insurance company pay insurance benefits if the Consorcio is already prepared to pay the benefits? Why should the owner of a building once again take out private insurance and once again pay premiums for this if he is already covered by the Consorcio and pays a surcharge for it (which is passed on to him by the private insurance company)? From the point of view of the private insurance, it would pay something and thus take on a financial burden that is already borne by the Consorcio. From the point of view of the policyholder, this would simply result in double insurance for which, furthermore, premiums would have to be paid, although coverage already exists which has to be paid compulsorily either way.
- As a result of the configuration of the legal provisions, owners of buildings must insure themselves with the Consorcio in any case and must pay a surcharge for this. There is no alternative. An offer by private insurances makes no sense. As a result, Spain thus has statutory insurance against natural hazards and natural disasters with the Consorcio, the state-owned enterprise, which still has a legal monopoly. The Consorcio's monopoly that was previously expressly referred to in the Estatuto Legal was given up superficially and ostensibly only, but not legally, when Spain joined the EU[96] (see p. 109 et seq. below on compatibility with EU law). In light of this circumstance, it is not surprising that the Consorcio is the only provider with respect to natural hazard insurance or that it has a "market share" of 100%[97]; in light of the legal monopoly, a market does not exist.

---

93 Art. 7, 18 Estatuto Legal; Art. 4 Reglamento del Seguro; Art. 1.1 and appendix I Resolución 27/11/2006.

94 Art. 18 no. 1 par. 2 Estatuto Legal; Consorcio, Diversity of Systems, p. 143.

95 Art. 18 no. 1 par. 2 Estatuto Legal.

96 The remarks in Consorcio, Diversity of Systems, p. 139, whereby the Consorcio lost its insurance monopoly in 1990, are therefore somewhat ambiguous. The Consorcio is not going to lose its monopoly based on the new Directive 2009/138/EC either, see footnotes 297-299 below.

97 Consorcio, Diversity of Systems, p. 140.

If need be, the Consorcio has a state guarantee in case it is unable to pay all insurance benefits[98]. Up to now, however, this state guarantee has not yet had to be utilised. To make provisions even for major loss events, the Consorcio has also built up fluctuation reserves[99].

The insured property and the insurance sum depend on the basic agreement of the private insurance; for buildings, this is usually a property insurance against fire and burglary. If replacement value coverage is included in the basic agreement, this also applies in the event the extraordinary risks named above occur[100]. Concluding a basic agreement is not obligatory but it is a prerequisite for coverage by the Consorcio. The latter is based on an existing basic agreement[101]. Protection by the Consorcio is thus circumvented if no basic agreement is concluded at all for buildings. However, due to the prevalence of fire insurance, this constellation nonetheless qualifies as compulsory insurance or, as in the present case, statutory insurance (see p. 29 et seq. above).

The following applies with respect to the amount of the surcharge and excess:

- The surcharge for the insurance of extraordinary risks is stipulated in tenths of a percent of the insurance capital of the basic agreement.
- Six risk categories were created according to the insured property. A separate rate applies to each risk category.
- Example: residential buildings rate 0.08‰, office buildings 0.12‰, commercial properties 0.18‰, industrial buildings 0.21‰[102].
- The different rates are based on many years of experience with losses. A gradation according to the *specific risk exposition* of the building based on natural hazards, e.g. the danger of flooding, does *not* take place, however. This ensures solidarity among policyholders.
- The excess for buildings is 7% of the replaceable loss, with no excess applied to residential buildings[103].

Finally, the importance of preventing losses caused by natural hazards was recognised even in Spain. Under the heading *"Other public responsibilities"*, Art. 16

98 Consorcio, Diversity of Systems, p. 145.

99 Art. 24 no. 2 Estatuto Legal; Real Decreto no 2013/1997 de 26 de diciembre 1997, de Regulación de las Provisiones Técnicas a dotar por el Consorcio de Compensación de Seguros [Royal decree no. 2013/1997 dated 26 December on regulating the equalisation fund to be formed by the Consorcio de Compensación de Seguros] (cited Real Decreto 26/12/1997).

100 Art. 8 no. 2 Estatuto Legal; Art. 5 Reglamento del Seguro.

101 Art. 3 no. 1, art. 6 no. 3 a), art. 7, art. 8 no. 3 Estatuto Legal; art. 4 Reglamento del Seguro.

102 Art. 13 Reglamento del Seguro; art. 1.1 and appendix 1 part 1 Resolución 27/11/2006.

103 Art. 8 no. 5 Estatuto Legal; Art. 9 Reglamento del Seguro.

Estatuto Legal therefore states another responsibility of the Consorcio: *"The development of plans and programmes to prevent and reduce cases of loss as well as the development of plans and programmes for campaigns and preventive measures to that effect."*

The Spanish insurance system corresponds to type 1 of the types of insurance systems presented.

2. Prevention of negative selection and maximum solidarity

The declared objective of the Spanish insurance system is, according to statements made by the Consorcio itself, to prevent negative selection and thus secure maximum solidarity among insured parties[104]:

"The Consorcio surcharge is the obligatory inclusion within the premium bill for all insurance policies within the modalities referred to both *when it is foreseen that the cover for extraordinary risks is to be assumed by the policy in question,* or *when the policy expressly does not assume such cover (in which case the Consorcio would assume responsibility for such cover).*

Its obligatory nature: The justification for the obligatory nature of this surcharge is based on the principles of COMPENSATION and SOLIDARITY *that preside in the Spanish system, without the application of which it would not be possible to support the natural* non-selection *of these risks. In effect, it is evident that if the payment of the surcharge was only required for the risks that one would voluntarily opt to have covered by the Consorcio, then only those with an appreciable degree of exposure to the risks would be willing to be included in the system, which would make it unviable from the start."*

Based on the statutory insurance system chosen, complete with legal monopoly, this objective has largely been achieved. Since all owners of buildings who have a basic agreement are insured with the Consorcio in any case and must pay the surcharge in any case, good risks are included in addition to bad risks. Equalisation of risks takes place.

---

104 According to a printout from the homepage of the Consorcio, www.consorseguros.es, heading Scope of Activity/Extraordinary risks/The surcharge and the tariff (as of 18 May 2009); likewise, Consorcio, Diversity of Systems, p. 143.

V. Switzerland

1. Overview—Coexistence of statutory insurance with integrated loss prevention/abatement and compulsory insurance

82% of all building values in Switzerland and thus the *vast majority* are protected against natural hazards by a system of *statutory insurance* with integrated loss abatement and loss prevention[105]. The corresponding task is carried out by the *Kantonale Gebäudeversicherungen* ("KGV"). The distinctiveness and the uniqueness of this system lies in the fact that the KGV, in addition to its function as a statutory insurance, is simultaneously also engaged directly in loss prevention and is furthermore responsible for fire services and national services (damage abatement). It is, therefore, not just an insurance system but a comprehensive system of *"securing and insuring"*. This system will be presented in detail on p. 56 et seq. below.

18% of all building values are insured against natural hazards by a system of *compulsory insurance.* This system, too, goes beyond the scope of the mere directive of an obligatory insurance since, to all intents and purposes, all essential elements of the insurance relationship are stipulated by the state. This system will be presented in detail on p. 64 et seq. below.

As a starting point, it should therefore be stated that in Switzerland, insurance against natural hazards is in no case left to market supply and much less to ad hoc state aid. Instead, this responsibility is borne by two systems that protect against natural hazards based on government provisions (primarily) by means of statutory insurance or by means of compulsory insurance.

2. The system of statutory insurance with integrated damage prevention and damage abatement

A. The Kantonale Gebäudeversicherungen (KGV) as solidary insurances

In 19 out of 26 cantons[106], statutory insurance, damage prevention and damage abatement is carried out by the KGV responsible for each canton. The KGV are independent public-law institutions. They are thus autonomous legal entities that are completely separated from the administration and that have their own assets. They

---

105 See p. 20 above.

106 Switzerland is a federal state. Its national territory consists of several member states, the cantons. Specifically, a KGV exists in the following cantons: Zurich (ZH), Berne (BE), Lucerne (LU), Nidwalden (NW), Glarus (GL), Zug (ZG), Fribourg (FR), Solothurn (SO), Basel-city (BS), Basel-county (BL), Schaffhausen (SH), Appenzell Ausserrhoden (AR), St. Gallen (SG), Grisons (GR), Aargau (AG), Thurgau (TG), Vaud (VD), Neuchâtel (NE), Jura (JU).

are based on cantonal law (building insurance law, "Gebäudeversicherungsgesetz", hereafter referred to as "GVG") which constitutes the KGV and lays down its powers and responsibilities[107].

According to the respective GVG, the owners of buildings must take out insurance against natural hazards, namely from the respective KGV. It is prohibited by law to take out insurance coverage from another insurance company. There is therefore not only compulsory insurance, i.e. obligatory insurance, but also *statutory insurance* that is administered by the KGV. For this purpose, the KGV have a *monopoly* for their respective cantonal region (see p. 83 et seq. below on the legal classification as well as the valuation of the monopoly in terms of constitutional and competition law). Insurance protection thus does *not* depend on a basic agreement (with another insurance company), nor is it based on extension of coverage, but obligatory insurance applies directly, and the exclusive and direct insurers are the KGV. In this sense, insurance protection is complete and cannot be circumvented—all building owners are included. On the other hand, the legal monopoly places the KGV under an obligation as well, for every building owner must be granted insurance protection. Thus, the KGV cannot, outside of exceptional cases defined by law (e.g. considerable loss risk due to completely insufficient maintenance), refuse anyone. The legal monopoly is therefore effective on both sides.

Since every building must be insured, the greatest possible risk collective is created. In addition, premiums (for details see below) are differentiated according to risk to a very minor degree only, if at all. This means that the KGV represent a *comprehensive insurance against natural hazards with maximum solidarity.* Good and bad risks are balanced. *The KGV are governed by the principle of solidarity.* Another consequence of the legal monopoly is that sales and marketing costs can be saved. Expensive continuous advertising and customer acquisition activities can be dispensed with. This cost saving can, on the one hand, be passed on to the policyholders by means of lower premiums. On the other hand, more financial means are available for loss prevention and loss payments.

Another prominent feature of the KGV is the fact that, in addition to their function as a statutory insurance, they are also responsible for prevention of damage caused by natural hazards and damage abatement, and are directly operational in these sectors, that is, they are not "only" involved in financial support. Thus, the KGV form unique *integrated welfare institutions* in this form. Solidary insurance is strengthened in the long term by the prevention and abatement of damage caused by natural hazards.

---

[107] Cf. for example the building insurance law of the Canton of Aargau dated 19 September 2009 (SAR 673.100), in particular § 2, as well as Loi concernant l'assurance des bâtiments et du mobilier contre l'incendie et les élements naturels (LAIEN) of the Canton of Waadt dated 17 November 1952 (RSV 963.41), in particular art. 1; VOGEL, Kommentar GV, p. 19, 22 et seq., 33.

This characteristic will be discussed in more detail in a separate paragraph on p. 56 et seq. below.

According to Swiss legal understanding, the KGV are therefore acting in the public interest. The corresponding legal ramifications will be discussed on p. 89 et seq.

From an actuarial point of view, the KGV are characterised by the following key features:

- *Insured natural hazards:* these are conclusively listed in the respective GVG, i.e. coverage is prespecified by law. With minor deviations, the hazards of storms, hail, floods, deluges, storm surges, avalanches, snow pressure, landslides and rock avalanches are insured[108] (for the sake of completeness only, it should be pointed out that the KGV also insure fire hazard under the statutory insurance system). This is therefore a case of comprehensive coverage in the sense of all-risk coverage.
- By law, the *replacement value* is insured in principle, i.e. the building owner cannot choose freely between replacement value and fair value, but instead replacement value insurance is compulsory[109]. In addition, the replacement value is, in principle, paid only if the building is rebuilt in principle. Replacement value insurance is therefore a reimbursement of costs insurance, as the costs of rebuilding are reimbursed[110].
- Premiums are differentiated in principle according to construction (sturdy, not sturdy, mixed) as well as according to type of use (residential, industry/commercial, agriculture, public buildings) (based on loss experience). In contrast, there is *no* gradation of risks based on the exposure to natural hazards. That is to say, the premium to be paid for a building situated on the bank of a river is no higher than that paid for a building on top of a hill. In addition, the largest KGV, as measured by the building insurance values (Gebäudeversicherung Canton of Zurich), charges a pure standard premium, i.e., no differentiation of premiums takes place, not even according to construction or type of use. The insurance tariff of each KGV is thus governed by the principle of solidarity as well; this fact is in part expressly stated in the GVG[111]. The same premium is not only the payment for insurance against natural hazards but also for fire hazard insurance. The premium is not split up between natural and fire hazards.

---

108 Cf. e.g. § 24 GVG LU as well as art.4 par. 1(e) GVG FR; Gerspach, Kommentar GV, p. 80 et seq.

109 Cf. e.g. § 5 GVG BS as well as § 7 and 8 GVG ZG; Rüegg, Kommentar GV, p. 153, 157 et seq.

110 For more details cf. Glaus, Kommentar GV, p. 253 et seq., p. 256.

111 As an example of a differentiated insurance tariff, cf. art. 21 GVG SH and § 19 et seq. Decree on the GVG SH. With respect to the standard premium of the KGV ZH, see § 43 GVG ZH. Regarding express reference to solidarity as a criterion of the insurance tariff, cf. art. 21 par. 1 GVG AR, art. 21 par. 1 GVG SH, art. 23 par. 2 and 3 GVG SG; Glaus/Kamber, Kommentar GV, p. 198 et seq.

- *Retentions* are stipulated by law as well, either with fixed amounts of CHF 200.00 to CHF 500.00 or by means of a percentage of the insured loss with a minimum and maximum amount (e.g. 10%, at least CHF 200.00, maximum CHF 2000.00). Only in few exceptions can an individual excess be agreed, which is routinely limited to 2% of the value of the building[112].
- The sum of insurance benefits is not limited, i.e. there is *no financial restriction* on the payments for which the KGV are liable to the building owner for losses, even in the event of a natural disaster. *No insurance limits* or maximum insurance or loss sums exist per catastrophic event or per year. Therefore, even in the event of a natural disaster with numerous aggrieved parties, the replacement value is guaranteed in each individual case, which is why this is also referred to in practice as "unlimited coverage". To be able to cope with this, the KGV created suitable tools with the IRV and the IRG early on.
- The KGV, being autonomous institutions under public law, are liable for their own obligations. Apart from an isolated case, there is no government guarantee[113].

As insurance companies that are organised under public law, the KGV are *not* subject to the Swiss *insurance supervision* executed by the FINMA[114] which can supervise insurance companies organised under private law only[115]. This, however, does not mean that there is no supervision. Instead, supervision is exercised by the relevant cantonal government and parliament, which procure the actuarial know-how required from third parties, if necessary.

The *insurance relationship* between the KGV and the policyholders is of *public-law* nature. In the statutory insurance sector, these are *not agreements* regulated by public law; instead they constitute a *decree relationship* since the KGV here act under sovereign jurisdiction (no coequality between the KGV and policyholders) and the GVG with their associated decrees do not leave any room for a contract[116]. All essential points of the insurance relationship are regulated by law and decree. There may be individual cases where, due to the leeway in configuration, a contract regulated

112 Cf. e.g. with respect to a fixed amount art. 35 GVG GR, with respect to a percentage § 33 GVG LU and with respect to an individual agreement art. 36a GVG SG; ANTHENIEN, Kommentar GV, p. 287 et seq.

113 The isolated case refers to the KGV NW, cf. art. 3 par. 3 GVG NW; VOGEL, Kommentar GV, p. 23, 33.

114 Financial Market Authority.

115 This is already apparent from the constitutional basis of art. 98 par. 3 BV, reference see footnote 123. Cf. VAG communication 1885, BBl 1885 I 101, in particular 107 as well as 629 (Ständerätliche Kommission report); VAG communication 1978, BBl 1976 II 873, in particular 879, 882, 883; VAG communication 2006, BBl 2003, 3789, in particular 3807 et seq.; WEBER/UMBACH, p. 42, 53; JACOBS, St. Gallen Kommentar BV, BV 98 N 14.

116 GERSPACH, Kommentar GV, p. 64; HÄFELIN/MÜLLER/UHLMANN, p. 225.

by public law can be assumed in spite of statutory insurance[117]. Depending on the canton, the insurance relationship comes into existence either automatically (at the start of construction) or it requires registration[118]. Should a building owner refuse to register, the KGV can stipulate the insurance relationship by decree, if necessary.

Individual KGV offer voluntary additional insurance policies or a movable property insurance outside of the monopoly. These types of coverage, however, are based on the GVG and the insurance relationship is regulated by public law. To be sure, these are *contracts* regulated by public law[119]. A contractual relationship under private law, even if the Insurance Contract Law (Versicherungsvertragsgesetz, ICL) is referred to as the supplemental applicable law in the GVG[120], doesnot exist as a rule. In this regard, private law is considered to be subsidiary public law[121].

The insurance system of KGV or statutory insurance practised primarily in Switzerland corresponds to type 1 of the types of insurance systems presented.

B. The Interkantonale Rückversicherungsverband (IRV) as an enhancement to the protection policy

The Interkantonale Rückversicherungsverband ("IRV") is a public-law corporation[122]. The bylaws of the IRV represent, as can be surmised from the company's name "Interkantonal" (intercantonal), an intercantonal agreement within the meaning of Art. 48 BV[123], or in other words, an interstate agreement[124]. Only public insurers—meaning, the KGV—can become members of the IRV. The General Assembly as the highest organ of the IRV consists of KGV delegates. The administrative board of the IRV must consist of management bodies from the KGV or members of the Swiss parliament[125].

According to Art. 2 of the bylaws, the purpose of the IRV is *"to assume the desired reinsurance coverage of its members by itself, to facilitate it, or to safeguard it (translation, original quote in German)."* From the institutional structure and the fact that only the KGV can be members of the IRV, it follows that the IRV is not an ordinary reinsurance. Instead, the IRV performs the same public task as the KGV by defraying losses caused by natural hazards above a certain amount, as part of

117 KOHLER/ZELGER, Kommentar GV, p. 352.
118 SCHAER, Kommentar GV, p. 144 et seq.
119 KOHLER/ZELGER, Kommentar GV, p. 353 (with respect to movable property insurance KGV GL according to GVG GL effective December 2009) and p. 394 et seq.
120 Federal law on insurance contracts of 2 April 1908 (ICL; SL 221.229.1).
121 HÄFELIN/MÜLLER/UHLMANN, p. 61 et seq.
122 Art. 1 par. 1 Bylaws IRV.
123 Federal Constitution of the Swiss Confederation dated 18 April 1999 (SL 101).
124 VOGEL, Kommentar GV, p. 35 et seq.
125 Art. 3 par. 1, art. 10, art. 12 par. 2 Bylaws IRV.

the comprehensive solidarity mechanism. Just like the KGV, the IRV is therefore tax-exempt because it serves a public purpose. Although it concludes "reinsurance contracts" with the KGV, it lacks an equalisation of risks according to the sector or a geographical equalisation of risks, since its exclusive contracting parties are the 19 KGV. The IRV purposely bears a cluster risk in the service of the KGV[126]. In the case of catastrophic events, the IRV even absorbs the losses within the framework of the IRG (see p. 52 et seq. below) without having received a premium for this so beforehand. Thus, according to Swiss law, this does not represent a reinsurance since at least one element of the notion of insurance is absent[127].

The enhancement to the protection policy by the IRV works in the following manner (between each KGV and the IRV, a corresponding "Reinsurance agreement against natural hazards" is concluded, where the contract clauses in the various contracts are always the same)[128]:

- Average losses caused by natural hazards should be defrayed by each KGV itself. The IRV and the KGV jointly define an appropriate retention concerning this matter.
- Above-average, that is, considerable losses should be defrayed primarily by the IRV, with a certain contribution by the KGV. For this purpose, the annual loss above the retention is distributed over several levels between the KGV and the IRV according to specific quota shares up to the large loss limit ("LLL"). The IRV here applies the technique of quota share reinsurance (which does not make a difference to the fact that the IRV is not a reinsurance). The IRV protects itself against the risks assumed by means of appropriate retrocession, i.e., reinsurance with third parties.
- Losses caused by natural disasters in the sense of extraordinarily high losses should be processed via the Interkantonale Risikogemeinschaft Elementar ("IRG"). Depending on the respective insurance capital (sum of building insurance values) and the claims history of a KGV, a LLL is defined for each KGV. Above this LLL, the loss is regulated through a special contract mechanism, the IRG. Within the IRG, a capacity of CHF 750 million is currently available.

Since an increase in high losses has been reported in recent years and the IRG were called to action in each case, the LLL of the KGV or the entry threshold for the IRG,

---

[126] Cf. IRV annual report 2008, p. 12.

[127] At least the element of orderliness of the business operation is absent (equalisation of risks according to the laws of statistics and the principle of large numbers), within the framework of the IRG, the element of the premium as well. Weber/Umbach, p. 54 et seq.; Streiff, p. 12 et seq.; FCR 114 Ib 244 E.4.a.

[128] Model Elementarschaden-Rückversicherungsvertrag IRV (effective 2009); IRV annual report 2004, p. 8 et seq.; IRV annual report 2008, p. 5, 14 et seq.

respectively, was markedly increased in 2005. This was, and is, to ensure that the IRG tool will not be applied until a disaster situation occurs[129].

C. The Interkantonale Risikogemeinschaft Elementar (IRG)—solidary defrayment of losses in the event of natural disasters

Even as early as 1996, the KGV reacted to climate change and the resulting increase in losses caused by natural hazards by introducing the IRG (see p. 20 et seq. above). The IRG is expressly designed for a disaster situation and should therefore take effect only in the case of extraordinarily high losses.

From a legal point of view, the IRG is neither a legal entity nor a mere company but instead a contract mechanism. It is based on a variety of contract clauses in the reinsurance contracts for losses caused by natural hazards between the KGV and the IRV. The IRG is essentially another layer, namely the uppermost layer for a disaster situation. The mechanism works as follows[130]:

- If the annual loss of a KGV exceeds the LLL, the IRG mechanism applies *to the loss exceeding the LLL.*
- Legally, the IRV is a debtor to the KGV for all payments made based on this mechanism.
- The initial CHF 25 million of this loss is borne by the IRV.
- The additional maximum CHF 500 million are paid by the IRV to the KGV affected by the catastrophic event. However, the IRV has *recourse* to *all* KGV for the entire amount paid. It therefore functions substantively as a *redistribution centre* between the KGV. *All* KGV pay a certain share of the loss to the IRV. The respective contribution by the KGV depends mainly on its insurance capital and also takes into account the LLL. From an economic point of view, all KGV therefore defray this part of the loss. This is a case of solidarity not among policyholders but *among the insurance carriers.* The KGV are defraying the loss solidarily. In practice, this means that a KGV will pay for a loss that did not occur in its area of responsibility. If, for example, the KGV BE is affected by losses due to flood that exceed the LLL by CHF 100 million, the KGV GR will absorb part of this loss (a share of CHF 75 million since the initial CHF 25 million are paid by the IRV), even if in the entire Canton of Grisons no flood damage or other losses caused by natural hazards were recorded. In a disaster situation therefore, all KGV, regardless of whether they are affected, defray the loss solidarily.

---

129 IRV annual report 2004, p. 10.

130 Model Elementarschaden-Rückversicherungsvertrag IRV (effective 2009); IRV annual report 2004, p. 8 et seq.; IRV annual report 2008, p. 14 et seq.

- The next maximum CHF 225 million are paid by the IRV in turn, which covers this share completely through retrocession.
- If the loss exceeded the capacity of the IRG, the affected KGV alone would again be called into action. In view of the total capacities based on various model calculations, however, this case should not come to pass. If the LLL amounts, for example, to CHF 350 million, the entire capacity of the statutory insurance system amounts to CHF 1.1 billion. This is sufficient to cover even high losses due to natural disasters.

The public statutory insurance system is shown in the diagram below[131]:

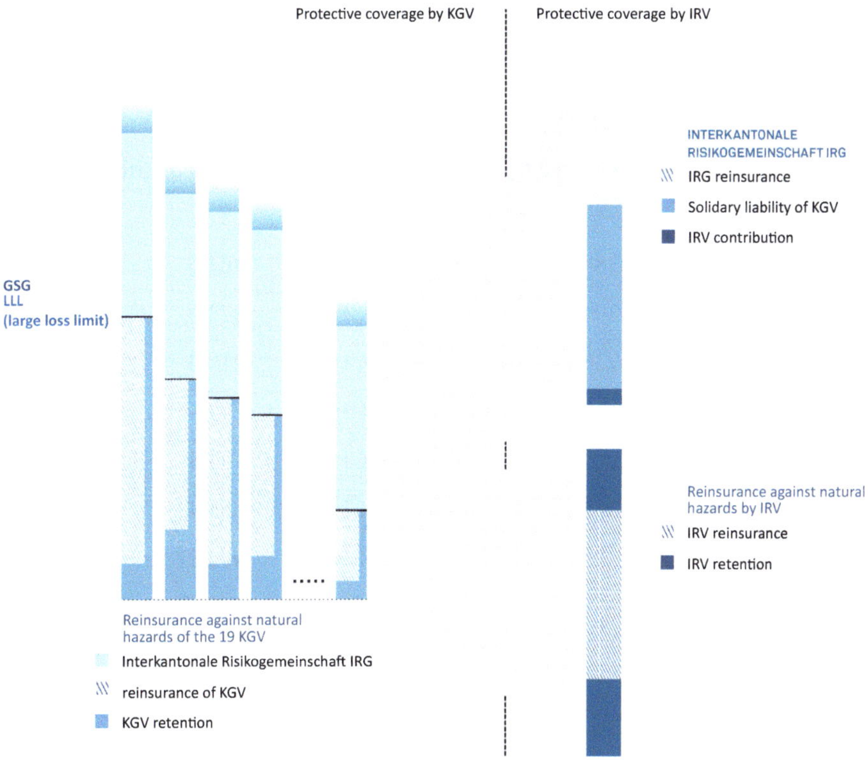

[131] Source of diagram: IRV annual report 2008, p. 15.

D. The protection policy has proven itself in practice against natural disasters

In 1999 and 2005, Switzerland was affected by natural disasters that had a considerable extent of damage. Thereby, the practicality in particular of the IRG, which had not been introduced until 1996, was put to the test. In both cases, the protection system consisting of KGV, IRV and IRG had produced proof of its practicality. The damage could be adjusted without any problems to speak of, and not even the smallest amount of ad hoc state aid would have had to be drawn on:

1999: in February, avalanches caused losses amounting to CHF 100 million; in May, floods caused losses amounting to CHF 200 million; in June and July, hail caused damage in the amount of CHF 100 million, and the winter gale "Lothar" cost as much as CHF 450 million in December. Add to this, smaller isolated loss events. All in all, building damage in the amount of CHF 957 million was recorded in the *cantons with KGV* insurance against natural hazards.

The protection system regulated by public law led to the following distribution of the claims burden[132]:

KGV		38%		CHF	364.6	million
IRV		39%		CHF	373.6	million
IRG		23%		CHF	219.4	million
		3%	IRV	CHF	29.4	million
		20%	KGV	CHF	190.0	million
KGV	total	58%		CHF	555.4	million
IRV	total	42%		CHF	402.2	million
Total				CHF	957.6	million

132 IRV annual report 1999, p. 5 et seq.; VKF annual report 1999, p. 8 et seq. Source diagram: IRV annual report 1999, p. 8, where according to information provided by the IRV in December 2009 the total final loss was CHF 957.6 million.

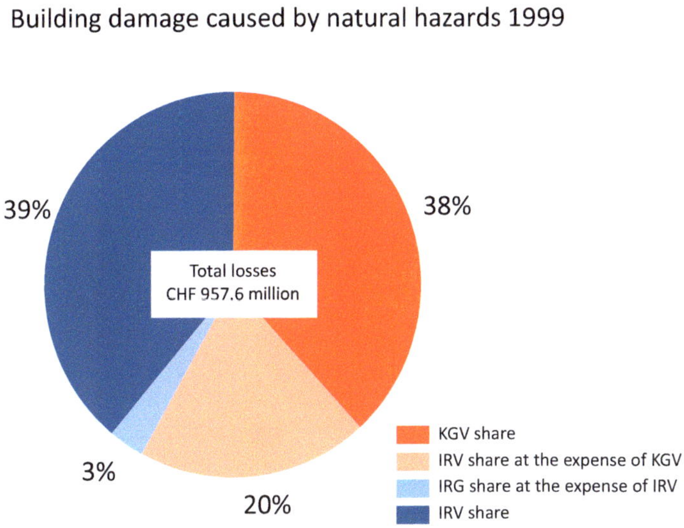

2005: this year was characterised by two disaster events in particular, namely a hail storm in July with losses of CHF 100 million and extensive flooding in August that caused losses of CHF 650 million. Together with other, smaller events, losses caused by natural hazards totalling CHF 870 million had to be overcome this year.

The protection system regulated by public law led to the following distribution of the claims burden[133]:

KGV		31%		CHF	270	million
IRV		28%		CHF	244	million
IRG		41%		CHF	356	million
		3%	IRV	CHF	26	million
		38%	KGV	CHF	330	million
KGV	total	31%		CHF	600	million
IRV	total	69%		CHF	270	million
Total				CHF	870	million

133 IRV annual report 2006, p. 7 et seq.; VKF annual report 2005, p. 22. Source of diagram: IRV Schadensstatistik 2009.

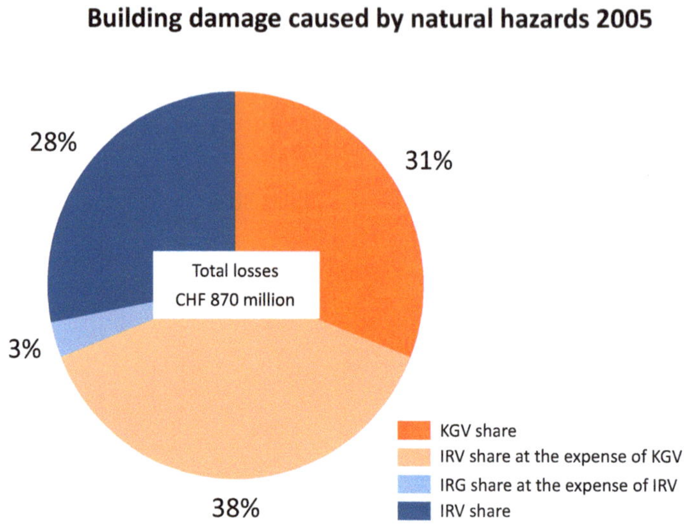

E. "Securing and insuring"—integrating the prevention of damage caused by natural hazards and damage abatement in the statutory insurance system

a) The basic idea—strengthening the insurance system and solidarity

Against the background of climate change and the increasing occurrence of losses caused by natural hazards that is looming, the KGV recognised the necessity of damage prevention early on[134]. To reduce the burden on the insurance system, or at least to stabilise it, preventive measures must be taken that prevent the accrual of damage or at least limit the extent of the damage.

The following basic idea is behind this:

- Prevention of losses caused by natural hazards decreases the costs of the KGV since fewer insurance benefits have to be paid due to less severe damage.
- Well-equipped and trained fire services (damage abatement) also accomplish this.

---

[134] As early as 1984, the KGV SG (building insurance organisation of the Canton of St. Gallen) had set up a fund for losses caused by natural hazards that contributed to the costs of property protection measures (e.g. sealing doors/windows, protective walls, elevating light shafts). In 1999, it issued a "Richtlinie Objektschutz gegen Naturgefahren" (guideline: property protection against natural hazards) demonstrating protective measures against gravitational natural hazards. In 2000, all KGV committed to the need for prevention and published an "Elementarschaden-Manifesto" (manifesto, regarding damage caused by natural hazards) containing a catalogue of specific measures to implement the prevention of losses caused by natural hazards (cf. VKF annual report 2000, p. 8).

- The KGV are not profit-oriented. Fewer costs lead to lower premiums for all building owners.
- As a result, firstly, the insurance system as a whole is strengthened further and, secondly, solidarity is strengthened. The insurance system is not weakened by excessive loss events and the favourable claims history allows affordable premiums to be offered to all policyholders, even those with bad risks, without the equalisation of risks by good risks being strained.
- This provides an incentive both for policyholders as well as for the KGV to pursue the prevention of damage caused by natural hazards. The KGV benefit from lower costs, the policyholders from lower premiums. For the same considerations, the KGV also have an incentive to invest in damage abatement (in particular infrastructure, equipment and training for the fire service).

The KGV public insurance system is therefore characterised not by its insurance function but by *"securing and insuring"*. The interaction can be graphed as follows:

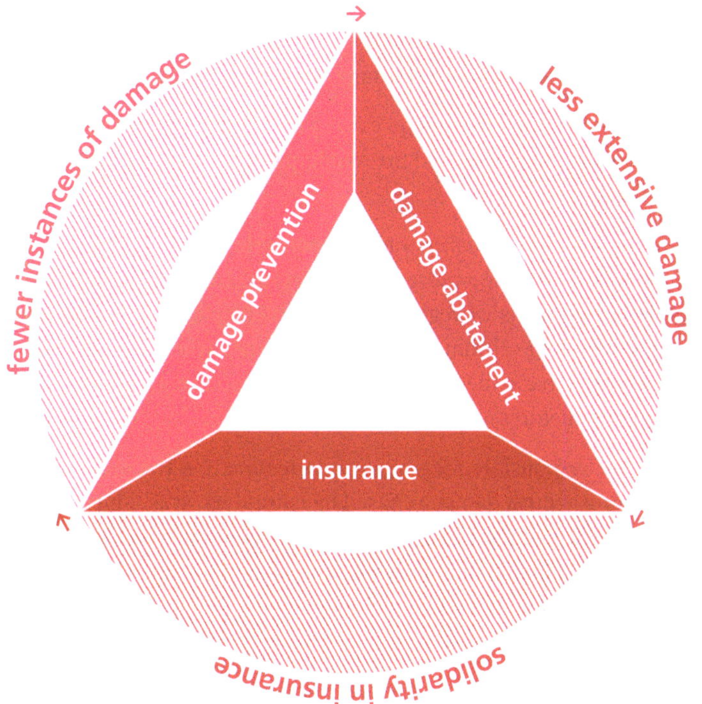

Due to the integration of insurance, damage prevention and damage abatement, numerous additional synergy effects can be achieved:

- Due to the determination of the building insurance value and the implementation of property protection measures, the KGV have excellent building data. They know the risks very well.
- These building data are of great use for the fire services in a damage event, especially natural disasters. The fire services know immediately what types of buildings are affected or threatened. Damage abatement can therefore be carried out in a more targeted and efficient manner.
- Conversely, insurance and prevention benefit from the specific experiences of the fire services during damage abatement. These in turn can be used to improve property protection or for other preventive measures.

All in all, due to these interactions and the synergy effects, the building owner benefits from a qualitatively superior and yet affordable overall performance of insurance, damage prevention and damage abatement. The service provision is extremely efficient. These correlations were recognised and acknowledged by the Swiss Federal Court ("FC" or "Federal Court") as well. With respect to the legal ramifications, we refer to p. 90 et seq. below.

b) Integrated prevention of damage caused by natural hazards

With respect to prevention of damage caused by natural hazards, a basic division of tasks exists between the canton and the municipalities on the one hand, and the KGV on the other hand[135]:

- The cantons and municipalities are responsible for the extensive and general protection against damage caused by natural hazards (area protection provisions). This is done by means of regional development planning, in particular hazard mapping and zoning (elimination of hazard zones where building is prohibited), as well as protective structures for the public good (e.g. general avalanche and flood control structures).
- The KGV are responsible for *property protection,* i.e. the protection of the individual building against damage caused by natural hazards. Thus, for example, art. 7 par. 2 sentence 1 of the Loi sur la protection contre les incendies et les dangers naturels of the Canton of Jura (law on fire prevention and protection against natu-

---

135 Cf. e.g. the guideline of the Canton of St. Gallen "Naturgefahren im Kanton St. Gallen, Leitfaden für Vorsorge und Schutz" (natural hazards in the Canton of St. Gallen, guideline for prevention and protection), Naturgefahrenkommission des Kantons St. Gallen, 2007 (cited Natural hazards guideline St. Gallen). The guideline can be accessed via www.gva.gvasg.ch, heading damage prevention (as of August 2009). Numerous sources of information, brochures and guidelines on the subject of prevention of damage caused by natural hazards can also be downloaded from www.kgvonline.ch, heading Dienste/Objektschutz (Services/Property protection) (as of August 2009).

ral hazards)[136]: *"Il [KGV JU] est l'autorité compétente en matière de protection des constructions contre les dangers naturels."* (It [KGV JU] is the responsible authority regarding the protection of buildings against natural hazards). This not only involves the protection of material assets but also personal protection.

The KGV pursue damage prevention using various specific measures:

- At the regional development planning level, hazard mapping, which is already partly completed, is supported by know-how, and personal and financial resources (hazard mapping should be completed by 2011).
- The KGV have published guidelines for property protection[137] against gravitational[138] and meteorological[139] natural hazards. These first inform of the effects of the respective natural hazard, in particular the typical types of damage, and contain detailed information on numerous protective construction measures. They are addressed to building owners, planners and architects. Thus, for example, in the "Wegleitung Objektschutz gegen gravitative Naturgefahren" (Guideline: property protection against gravitational natural hazards), several hazard scenarios with respect to flooding are illustrated in detail (static flood, dynamic flood, etc.). Then, they describe how the various effects of floods can be determined. Subsequently, an overview of possible protective measures is first presented (wetness precautions, waterproofing, screening). Based on this, numerous individual measures are described in detail (e.g. appropriate positioning/installation of supply facilities such as power supply, oil tanks, etc.; backwater protection canalisation, waterproofing/reinforcing doors/windows etc.; elevating light shafts etc.; sealing the cladding/foundation; elevating the building e.g. on stilts; building protective barriers and walls), where their suitability for existing buildings or new buildings as well as the combination of different measures is shown. The guidelines represent practice-oriented work tools that show build-

136 Loi sur la protection contre les incendies et les dangers naturels of the Canton of Jura of 21 November 2007, effective since 1 January 2009 (RSJU 871.1).

137 Guideline Objektschutz gegen gravitative Naturgefahren (property protection against gravitational natural hazards); Guideline Objektschutz gegen meteorologische Naturgefahren (property protection against meteorological natural hazards). The guidelines can be downloaded from www.kgvonline.ch or www.vkf.ch, each under the heading Dienste/Objektschutz (services/Property protection) (as of August 2009).

138 Avalanches, floods, landslides, debris flows, rock fall, block fall, rock avalanche, rock slide, ice slides. The main drive of these natural hazards is gravitational force, and the area affected is usually limited by the topography. Thus, these hazards do not occur everywhere in Switzerland but are associated with a locality. Cf. Guideline Objektschutz gegen meteorologische Naturgefahren (property protection against meteorological natural hazards), p. 11.

139 Wind, hail, rain, snow, lightning. These are direct hazards arising from short-term weather phenomena that cannot be spatially avoided. Thus, they occur everywhere in Switzerland. Cf. guideline Objektschutz gegen meteorologische Naturgefahren (property protection against meteorological natural hazards), p. 11.

ing owners, planners and architects in concrete terms what prevention measures can be implemented and how. In addition, the guidelines are tips that should be heeded rather than being merely voluntary and, depending on the canton, they are issued with a certain binding nature by reference to legislation[140] and/or actuarial measures (see below).

- Depending on the canton, the KGV participate in construction permit procedures. Within this framework, property protection measures can be imposed for new buildings. In the Canton of St. Gallen, for example, construction projects in areas with considerable hazards require an application for a derogation, which can only be issued after evaluation by the KGV SG. For construction projects in areas with an average hazard rating, the construction permit will be issued only if the hazard can be reduced to a sustainable level through property protection measures. The latter are mandatory in the construction permit with the involvement of the KGV SG. In low-hazard or residual-hazard areas, property protection measures are stipulated for specific buildings, e.g. buildings for which large crowds of people can be expected[141]. Furthermore, the KGV SG has a fund for averting damage caused by natural hazards. Construction projects can be supported with contributions from this fund if the new building must be constructed in an area that is at risk due to crucial public interests[142]. In another example, the Canton of Jura, the KGV is likewise involved in the building permit procedure and can stipulate prevention measures[143]: *"L'autorité compétente* [KGV JU] *peut prescrire à des collectivités publiques ou à des personnes privées des mesures particulières de protection des constructions contre les dangers naturels, tels que murs, barrage, digues, canalisations, travaux de stabilisation ou de consolidation, etc."* (The responsible authority [KGV JU] can stipulate that public facilities or private persons take measures to protect buildings from natural hazards, such as walls, barriers, dikes, canalisation, stabilisation or reinforcement work, etc.).

---

140 Cf. e.g. Art. 6 par. 1 Loi sur la protection contre les incendies et les dangers naturels of the Canton of Jura, according to which the government can declare technical regulations by specialised organisations that are recognised in the sector of prevention to be obligatory. Art. 5 Ordonnance sur la protection contre les incendies et les dangers naturels et sur le ramonage of the Canton of Jura dated 18 November 2008 (decree on fire prevention and protection against natural hazards as well as the chimney sweep business, RSJU 871.11) records that the binding regulations according to art. 6 par. 1 of the Act are listed in annex I of the decree. Finally, the guidelines Objektschutz gegen gravitative und meteorologische Naturgefahren (property protection against gravitational and meterological natural hazards) are listed under no. II/1.3 annex 1. Cf. also art. 18 par. 1 of the above-mentioned Act.

141 Guideline Naturgefahren SG (natural hazards, St. Gallen), p. 26 et seq., with reference to the pertinent statutory provisions.

142 Art. 1 par. 2 Decree on the contributions from the fund for averting damage caused by natural hazards for the canton of St. Gallen dated 5 July 1983 (sGS 873.12, hereafter referred to as "Fund SG decree").

143 Art. 17 Loi sur la protection contre les incendies et les dangers naturels des Kantons Jura. Cf. also Art. 9 and Art. 10 par. 1 of this Act.

- For *existing* buildings, property protection measures are, depending on the canton, financially supported by the KGV and/or can be mandated by them. We again refer to the provisions of the Cantons of St. Gallen and Jura as examples. In the Canton of St. Gallen, contributions are awarded from the above-mentioned fund if the building is completely or largely protected from imminent damage caused by natural hazards by the protective measure[144] (a maximum of 50% of the costs of the protective measure are assumed, or 70% if the applicant is suffering from financial hardship). According to the regulation of the Canton of Jura, property protection measures are prescribed for conversions, extensions or changes of purpose/utilisation of buildings as well as in cases where personal protection is of particular importance[145].
- With *actuarial measures* (exclusion from insurance and coverage, provisos), the KGV can work towards appropriate designs and building materials being used, property protection measures being implemented and the building being maintained properly at least (poorly maintained or even neglected buildings are regularly more susceptible to damage caused by natural hazards). As an example, we refer to the regulation in the Canton of Zurich[146]. If it turns out in a loss event caused by hail that, for example, the winter garden had been built using completely unsuitable glass material, the KGV ZH can exclude coverage due to unprofessional and unsound building construction. Hail is consistently considered a foreseeable hazard. The damage was thus foreseeable and is not covered. In addition, a building can be partially excluded from insurance from the start if, in view of natural hazards, completely unsuitable designs or materials were used (for example, a completely unsuitable glass façade, for aesthetic reasons) (the KGV must warn the owner of the building beforehand). Thereby, solidarity is not unnecessarily strained, and the owner of the building can take out insurance coverage for the excluded part of the building from a private insurance company. In any case, however, it is assumed that the building owner has access to information about the suitability or unsuitability of designs and construction materials and construction components. Furthermore, actuarial measures are another means of enforcing property protection measures. Actuarial measures on the whole have a preventive effect since the building owner is more or less forced to use the appropriate design and materials/building components if he does not want to risk being excluded from insurance or coverage.
- With regard to hail, a hazard that occurs frequently in Switzerland, and with devastating losses, the VKF developed and introduced the *Schweizerische Ha-*

---

144 Art. 1 par. 1 "Fund SG decree".

145 Art. 10 par. 1 Loi sur la protection contre les incendies et les dangers naturels of the Canton of Jura.

146 Cf. § 20 no. 3, § 12 GVG ZH. Reference see footnote 151.

gelschutzregister (Swiss hail protection registry) *("HSR")* in 2008[147]. It contains a classification of building materials and building components according to resistance to hail (Hagelwiderstand, "HW") with levels 1–5 (5 being the strongest HW) and, as an application registry, ensures that suitable products are used (the HSR does not stipulate how building materials and building components are to be manufactured but provides information regarding what applications they are suitable for. Thus, for glass roofs, for example, various products are classified according to their HW. From this, it follows which products are suitable for a winter garden, for example, and which are not). The HSR may serve as a discussion starter for actuarial measures even now. It will definitively be in effect as from 1 January 2010 when the KGV recommends the HSR as a "tool and decision-making aid" to builders and architects. It therefore has a preventive effect. In the medium term, the HSR is scheduled to be anchored in the set of regulations of the Swiss Association of Engineers and Architects (Schweizerischer Ingenieur- und Architektenverein, SIA). The SIA standards are often recognised as the rules of engineering which means that non-compliance with these rules represents a breach of the duty of care under private law (which can, for example, amount to a breach of the Architect's Contract). In the long term, an obligation under public law is also conceivable in the sense that it becomes mandatory to use the appropriate products only (analogous to fire prevention). The Canton of Jura has now declared the HSR to be binding by reference to the law[148].

c) Integrated damage abatement

In addition to damage prevention, the KGV also performs certain tasks of damage abatement. Depending on the canton, damage abatement is integrated more or less deeply into the KGV. The positive effects of this integration have already been pointed out.

Apart from firefighting, the fire services are also responsible for abating damage caused by natural hazards (e.g. pumping out flooded cellars; setting up mobile flood barriers)[149]. The municipality is usually the provider of fire services. In this sector, however, the KGV performs various tasks, as shown below briefly by reference to the KGV ZH ("GVZ")[150] and FR:

---

147 The HSR can be accessed via www.hagelregister.ch or www.kgvonline.ch, heading Dienste/Hagelschutzregister (services, hail protection registry) (as of August 2009).

148 Annex 1 no. II/1.3 Decree on fire prevention and protection against natural hazards as well as the chimney sweep business of the Canton of Jura.

149 Cf. § 16a par. 1(a) fire police and firemanship Act of the Canton of Zurich of 24 September 1978 (LS 861.1, hereafter referred to as "Fire Services Act ZH").

150 Gebäudeversicherung Kanton Zurich (building insurance, Canton of Zurich).

- In accordance with § 2 par. 2 GVG ZH[151], *[the GVZ] provides the fire police and firemanship based on special statutory provisions, insofar as these tasks are incumbent upon state entities. Additional sectors of personal and material asset protection can be transferred to it (translation, original quote in German)."*
- The GVZ has a direct influence on the structure of the fire services by being able to designate municipalities and professional fire services as bases for regional assistance in the case of special or major events. Professional fire services can be set up only in accordance with the GVZ[152].
- The GVZ determines the performance targets with respect to the operational readiness of the fire services[153].
- The institution *"cantonal fire services",* which monitors the organisation, alerting, training and equipping of the (communal) fire services, *"is exercised by the GVZ"*[154].
- The municipality can specify how its communal fire services are organised only in accordance with the GVZ. In addition, the GVZ appoints cantonal operations managers. The fire services are trained according to instructions given by the GVZ, and the GVZ itself shall carry out the basic courses and the promotion courses, among other things[155].
- Infrastructure and training of the fire services shall be financed to a considerable extent by the GVZ[156].
- In the Canton of Fribourg, the law pertaining to the fire police and the protection against damage caused by natural hazards aims at *"the protection of buildings against fire hazards and other damage caused by natural hazards (translation, original quote in German)"*[157]. The KGV FR is one of the executive bodies of this *law, in particular the executive body for all issues regarding the prevention and abatement of damage caused by fire and natural hazards*[158].
- Here, too, the municipality is in charge, in principle, of setting up fire services. However, the KGV FR can set up fire service bases[159]. Fire service commanders,

151 Building insurance Act of the Canton of Zurich of 2 March 1975 (LS 862.1).
152 § 19, 20 Fire Services Act ZH.
153 § 22 Fire Services Act ZH.
154 § 24, 24a Fire Services Act ZH.
155 § 3, 10, 13 Fire Services Decree of the Canton of Zurich dated 22 April 2009 (LS 861.2; hereafter referred to as "Fire Services Decree ZH").
156 Cf. among other things § 31 Fire Services Act ZH; § 4, 14 Fire Services Decree ZH.
157 Law pertaining to the fire police and the protection against damage caused by natural hazards of the Canton of Fribourg dated 12 November 1964, Art. 1 par. 1 (SGF 731.0.1, hereafter referred to as "Fire Police Act FR").
158 Art. 3(e), art. 8(c) Fire Police Act FR.
159 Art. 33, 34 Fire Police Act FR.

even from the communal fire services, can only be appointed in accordance with the KGV FR. In addition, the latter is in charge of training[160].

F. Prevention of negative selection and moral hazard

The KGV statutory insurance system *prevents negative selection.* Due to the legal monopoly, all risks, including good risks, are incorporated in the risk collective. Due to the *maximum solidarity,* the *cross-subsidisation* of bad risks by good risks, and thus the solidary insurance, is not strained. For good risks, there is no reason to look for alternatives. Even these building owners will benefit from the incorporation of all risks. An accumulation of bad risks does not occur in the KGV from the outset. The *financial stability* of the insurance system is thus *guaranteed* even in years with a high claims burden.

Furthermore, *moral hazard* is *prevented.* By means of the partially binding and/or financially supported property protection measures, as well as the actuarial measures, owners of buildings are either forced to practice prevention of damage caused by natural hazards, or else there is at least a strong incentive to take preventive measures. Without individual provisions, building owners will not receive a building permit at all, or they risk not receiving insurance benefits in a loss event.

Owing to the KGV statutory insurance system, two substantial negative incentives are thus neutralised.

3. The compulsory insurance system

A. Compulsory insurance in the sense of obligatory extension of coverage

In 7 cantons[161], losses caused by natural hazards are not insured by a KGV under public law but by private insurance companies. If the building owner has taken out a fire insurance for buildings, coverage is extended by law to include natural hazards (statutory extension of coverage). Thus, the prerequisite for insurance against natural hazards is the conclusion of a basic fire insurance agreement. In the Cantons of UR, SZ and OW, the conclusion of a basic agreement is obligatory[162]. In the Cantons of GE, TI, AI and VS, the conclusion of a basic agreement is voluntary, so that the in-

---

160 Art. 458, 462, 462a Decree pertaining to the fire police and protection against damage caused by natural hazards of the Canton of Fribourg dated 28 December 1965 (SGF 731.0.11).

161 Cantons of Geneva (GE), Uri (UR), Schwyz (SZ), Ticino (TI), Appenzell Innerrhoden (AI), Valais (VS), Obwalden (OW), also referred to as "GUSTAVO cantons". Cf. WANNER, Kommentar GV, p. 5.

162 Art. 3 Law on the obligatory building insurance of the Canton of Uri dated 7 March 1993 (RB 40.1402); § 77 Introductory law to the Swiss Civil Code dated 14 September 1978 of the Canton of Schwyz (SRSZ 210.100); art. 21 Law on the official appraisal of properties and the lien on property dated 26 October 2006 of the Canton of Obwalden (GDB 213.7). In the above-mentioned provisions, insurance against natural hazards is declared obligatory as well. The obligatory insurance to this ef-

surance against natural hazards aimed at by the legislature can be undermined. However, since the insurance density for fire insurance is very high, this, too, is referred to as compulsory insurance or obligatory insurance.

From the legislature's point of view, extensive insurance against natural hazards is in the public interest[163]. As shown above in 19 cantons, the cantons themselves have respected this by creating the KGV statutory insurance system. The cantonal legislature took action early on, and the KGV are based on cantonal law. In 7 cantons, however, the federal legislature provided insurance against natural hazards. In these cantons, it is based on federal law. The Swiss VAG[164] and the associated supervision ordinance (Aufsichtsverordnung, "SO")[165] regulate the insurance against natural hazards as follows:

- The *scope of coverage* is statutory, uniform for everyone and includes[166]: deluge, flooding, storm, hail, avalanches, snow pressure, rock avalanche, rock fall and landslides.
- By law, the *replacement value* ("full value")[167] is insured.
- Private insurance companies must develop a premium calculation formula, on the basis of which a premium rate for a *standard premium is submitted.* Both are approved by the FINMA[168], the Swiss supervisory authority for private insurance companies[169].
- Even the ceiling for *retentions* is statutory, namely 10% of the insured losses for residential and agricultural buildings, at least CHF 1,000.00 and at most CHF 10,000.00, and for other buildings (in particular commercial and industrial buildings) 10% of insured losses, at least CHF 2,500.00 and at most CHF 50,000.00[170].
- The sum of insurance benefits in all is limited by law, i.e. there is an *insurance limit or liability limit* for private insurers for damage to buildings. Thus, insur-

fect, however, already arises from federal law, namely from the VAG and the SO (see footnotes 164 and 165 below).

163 Cf. Communication "direct insurance agreement and SchVG", BBl 1991 IV 24 et seq. Cf. also WEBER/UHLMANN, p. 168, who describe the ES pool as a "Sozialwerk" (social institution); this will have to apply to the KGV all the more.

164 Federal law regarding the supervision of insurance companies (Insurance Supervision Law/Versicherungsaufsichtsgesetz) dated 17 December 2004 (SL 961.01).

165 Ordinance on the supervision of private insurance companies dated 9 November 2005 (SL 961.011).

166 Art. 33 par. 2 and par. 4(b) VAG; Art. 173 par. 1 SO. The statutory extension of fire insurance coverage is provided for in art. 33 par. 1 VAG and art. 171 par. 1 SO.

167 Art. 171 par. 1 SO.

168 Finanzmarktaufsicht (financial market supervisory authority).

169 Art. 33 par. 2, par. 3, par. 4(a) VAG; art. 177, 178 SO.

170 Art. 33 par. 4(b) VAG; art. 175 par. 1(d) SO.

ance benefits paid out by all private insurance companies combined are limited to CHF 25 million per event for an individual policyholder, and to CHF 1 billion[171] for an insured event (regardless of the number of policyholders). It should be noted that these limits do not apply per insurance company, i.e., the capacity does *not* amount to CHF 25 million or 1 billion, resp., per insurance company, but instead the capacity of the entire *private insurance industry* that covers losses caused by natural hazards is limited to these amounts. If the losses exceed the above-mentioned limits, all insurance benefits are cut proportionally, such that the limits are observed[172].

The type of compulsory insurance system described here corresponds to type 2 of the types of insurance systems presented.

B. The Swiss insurance pool for damage caused by natural hazards for balancing of risks and prevention of negative selection

15 Private insurance companies that dominate 95% of the market in the GUSTAVO cantons (as of 2009) pass the risks on to the Swiss insurance pool for damage caused by natural hazards (Schweizerischer Elementarschaden-Pool, “ES pool”)[173]. This is a simple partnership pursuant to art. 530 et seq. of the Swiss Code of Obligations (Schweizerisches Obligationenrecht, OR). However, it is the private insurances exclusively, not the ES pool, that have an obligation towards the policyholders; policyholders have claims only against private insurance companies, not against the ES pool. Broadly speaking, this works as shown below[174]:

- Each associated private insurance (pool member) itself defrays 20% of the compensation.
- 80% of the compensation is pooled.
- Each pool member bears a share in the pooled losses according to its pool percentage share. For buildings, this corresponds to the quotient of the average fire insurance capital of the pool member concerned and the entire averaged fire insurance capital of all pool members. The pool percentage shares are recalculated each year at an annual rate.

---

171 Until the end of 2006, this limit was set at CHF 250 million which turned out to be much too low in view of the increasing damage due to natural hazards. The limit was therefore quadrupled as of 1January 2007 and thus set to CHF 1 billion.

172 Art. 33 par. 4(b) VAG; art. 176 SO.

173 Information on the ES pool can be obtained from the homepage of the Schweizerischer Versicherungsverband (Swiss insurance association, SVV), www.svv.ch.

174 Pool agreement of the Swiss insurance pool for damage caused by natural hazards (as of 30 November 2007, valid for losses as from 1 January 2008), in particular art. 2, 5 and 6; Handelszeitung dated 5 July 2006.

- Pool membership is voluntary. Should the ES pool threaten to fall apart, however, which would jeopardise insurance against natural hazards, the Federal Council can mandate pool membership by law[175].

Pooling compensation payments and losses is intended to balance risks and thus achieve a certain degree of *solidarity* between good and bad risks, that is, among policyholders. This ultimately also allows all risks to be insured in exchange for payment of a standard premium. This is also where solidarity among policyholders comes in.

Furthermore, *negative selection* is largely prevented by compulsory insurance. Even good risks must take out insurance. This is compulsory, at least in the cantons that have obligatory insurance for the basic agreement (fire insurance). In cantons without this obligation, statutory extension of coverage should also occur in most cases since fire insurance is widespread. An accumulation of bad risks is thus prevented for the most part. Should individual private insurance companies nonetheless record an accumulation of bad risks in their portfolio, the ES pool will allow them a certain degree of cushioning which allows them to secure their financial equilibrium. Thus, solidarity is also created among pool members.

C. The compulsory insurance system as a statutory market and price system in accordance with antitrust law

How should the compulsory insurance system be classified? Does it represent a private or a market-based solution?

According to art. 3 par. 1(a) of the Swiss antitrust act (Kartellgesetz, "KG")[176], provisions are reserved *"if they do not allow competition on a market for certain goods or services, in particular provisions that establish a state market or price system (translation, original quote in German)."* Such competition-limiting provisions take precedence over the KG, i.e. the KG is essentially not applicable[177]. They are designed precisely to exclude the market for higher-weighted interests:

A state market or price system exists if

- The legislature assumed a market failure when the (competition-limiting) provisions were adopted (whether market failure actually exists is immaterial and does not even need to be examined)[178]. Evidence that it is the express will of the legislature to exclude competition is not required. It is sufficient for the provisions

---

175 Art. 33 par. 5(b) VAG.

176 Federal law on cartels and other restrictions on competition dated 6 October 1995 (SL 251).

177 The competition commission can only make recommendations according to art. 45 KG; it cannot perform any inspections, much less adopt provisions.

178 ZÄCH, Kartellrecht, p. 133.

to be incompatible with competition, from which it follows that the sector is not subject to competition law[179].

- the important economic parameters, that is to say, the economically fundamental elements of a market, are decisively laid down by public law. Such parameters are, e.g., pricing, production, distribution, offer and acceptance[180].

These prerequisites are met in the present case:

- When the provision on insurance against natural hazards was adopted, the legislature assumed that, without such a provision, extensive insurance protection at affordable premiums could not be implemented. In addition, the need for the legislation to intervene for reasons of social policies and national economics, and the threat to the system posed by market liberalisation, was repeatedly pointed out[181]. The legislation thus assumed market failure, namely that the market cannot guarantee a satisfactory offer for insurance against natural hazards.
- Both the extent of coverage as well as the premium rate are statutory, namely within the meaning of uniform coverage and standard premium. In addition, the insurer is subject to the obligation to offer insurance. Even the insured value, the excesses and thus the insurance benefits are statutory. Thus, the significant economic parameters are decisively laid down by public law, which means that competition and market should be excluded.

Accordingly, the compulsory insurance system in the GUSTAVO cantons represents a statutory market and price system within the meaning of the KG. Even the competition commission concluded in 2003 that the federal regulations regarding insurance against natural hazards permit neither price nor quality competition among private insurers and that therefore the KG does not apply[182].

Although compulsory insurance is implemented by private insurance companies, this does not therefore constitute a private-law or market-based solution. Instead, the private insurance companies fulfil a public duty when implementing insurance against natural hazards and are an administrative agency under private law, comparable to health insurances organised under private law which implement social health insurance[183].

---

179 FCR 129 II 497, in particular 515 = Pra 94 (2005) no. 39 p. 290, in particular p. 304 et seq. BORER, p. 91 et seq.

180 WEBER, SIWR V/2, p. 47; ZÄCH, Kartellrecht, p. 113 et seq.

181 Communication "direct insurance agreement and SchVG" (see footnote 163 above), 24 et seq. on art. 38a of the old VAG which was adopted without change into the current VAG as art. 33.

182 Recht und Politik des Wettbewerbs (RPW) 2003, p. 743.

183 Cf. HÄFELIN/MÜLLER/UHLMANN, p. 318 et seq.

D. Comparison between the KGV statutory insurance system and compulsory insurance

The following essential differences can be found between the two systems:

- Neither damage prevention nor damage abatement is integrated into the private insurances in the compulsory insurance system. In the cantons that do not have the KGV, there is not connection between insurance and damage prevention or damage abatement. The latter is performed by the responsible local authorities alone.
- In contrast to the KGV, private insurances do not have a legal monopoly. Thus, various providers operate in the compulsory insurance system, albeit not on a market but within the framework of a state market and price system. As is the case in the cantons that have the KGV, both offer and price are statutory; what differs is that only a single provider exists in the statutory insurance system. In contrast to private insurances, the KGV can therefore save sales and marketing costs.

The differences mean that in the statutory insurance system, due to the interaction and synergy effects between insurance and prevention or damage abatement, the claims burden is lower. Furthermore, the KGV do not incur sales and marketing costs due to the monopoly. The premium level in the statutory insurance system is therefore more favourable than in the compulsory insurance system (see p. 90 et seq., 103 and footnote 275 below).

VI. USA

1. Insurance option based on market supply with state support

A. National Flood Insurance Program (NFIP)

In the USA, the natural hazards of deluge/flood, storm, hurricane and earthquakes are particularly relevant. An overview of the first two hazards is provided below.

The American system of dealing with building damage as the result of flooding is an interaction between the private insurance market and state authorities, which is outlined under the heading of National Flood Insurance Program ("NFIP"). It became necessary after private insurances had almost completely withdrawn from the insurance market for flood risks, following extensive floods and hurricanes in the late 1960s[184]. Insurance contracts are usually underwritten by a private insurance

---

[184] The NFIP was founded in 1968 and is essentially based on the 1973 Flood Disaster Protection Act as well as the 1994 and 2004 Flood Insurance Reform Acts. Cf. FEMA leaflet, Myths and Facts about

company which is also responsible for claims settlement. The risk, however, is borne by the Federal Emergency Management Agency ("FEMA")[185], a state institution at federal level that reports to the US Department of Homeland Security (DHS). FEMA takes on a function that is similar to that of a reinsurance. Private insurance companies transfer those premium revenues that exceed the claims payments to FEMA, which uses them to support the National Flood Insurance Fund ("NFIF"). If, however, the claims payments exceed the premium revenues, FEMA will reimburse the private insurance company for the excess loss[186]. The necessary resources are taken from the NFIF.

The decisive issue now is how access to the NFIP is regulated[187]:

- The building owner will *have access to insurance protection only* if the *community* where the building is located has *joined* the NFIP. There is *no obligation whatsoever for a community to join.* The latter is thus voluntary even if certain mechanisms exert pressure in favour of joining. For a community to be able to join at all, it must already either have issued regional development planning and construction regulations or must pass such regulations within a useful period of time. It must have a "floodplain management": *"Floodplain management is the operation of a community program of corrective and preventive measures for reducing flood damage. These measures take a variety of forms and generally include requirements for zoning, subdivision or building, and special-purpose floodplain ordinances."*[188].
- Even if the community has joined, there is still *no compulsory insurance* for the *building owner.* It is up to the building owner alone whether he wants to take out insurance. An exception exists for building owners whose buildings are located in a Special Flood Hazard Area ("SFHA")—a zone that is particularly exposed to

---

the National Flood Insurance Program, which can be downloaded from www.fema.gov via the link "NFIP", heading "Flood Insurance" (as of 20 October 2009); Consorcio, Diversity of Systems, p. 182; Kunreuther/Michel-Kerjan, in: Richardson/Gordon/Moore II, p. 15 et seq.

185 The purpose of FEMA is *"to prepare for, protect against, respond to, recover from, and mitigate all hazards"*, with the triggering event being a natural disaster or a terrorist act, see www.fema.gov, heading "About FEMA" (as of 20 October 2009). FEMA as an institution was founded in 1979 and, as a consequence of Hurricane Katrina, was adapted to the new requirements by the Post Katrina Emergency Management Reform Act. FEMA's headquarters are in Washington, D.C. Furthermore, FEMA has more than 10 regional offices which are distributed all over the country. Cf. the FEMA leaflet Prepared. Responsive. Committed., which can be downloaded from www.fema.gov, heading "About FEMA" (as of 20 October 2009).

186 Consorcio, Diversity of Systems, p. 182, 187, 190.

187 FEMA leaflet, Myths and Facts about the National Flood Insurance Program, which can be downloaded from www.fema.gov via the link "NFIP", heading "Flood Insurance" (as of 20 October 2009); Kunreuther/Michel-Kerjan, in: Richardson/Gordon/Moore II, p. 15 et seq.; Consorcio, Diversity of Systems, p. 182 et seq., 184 et seq.

188 Cf. www.fema.gov/plan/prevent/floodplain (as of 20 October 2009).

risks. If these owners have taken out a mortgage loan that is granted or secured by federal institutions, compulsory insurance exists. However, even this does not change the fact that access to insurance protection and thus compulsory insurance depends, even in this case, on whether the community in whose area a SFHA is located has joined the NFIP. The USA are attempting to exert pressure at federal level to enforce compulsory insurance by excluding communities with SFHA that have not joined the NFIP from ad hoc state aid in the event of a natural disaster. In addition, no federally funded mortgage loans can be taken out in these areas. Ultimately, compulsory insurance has a very weak footing, as a building owner in a SFHA who has not taken out a federally funded mortgage loan is not subject to the compulsory insurance, nevermind the fact that the NFIP as a whole is circumvented if the community does not join.

- The US system regarding floods is therefore characterised by *double voluntariness.* It is left to the community to join the NFIP. Even if a community has joined, it is voluntary for the building owner to take out insurance. Hence, insurance density is very low. In terms of single-family households, it is about 49% on average within the SFHA but only just 1% outside of the SFHA[189].

From an actuarial point of view, the NFIP is characterised as follows:

- Various types of policies are offered under the NFIP. For single-family homes, the Standard Form Insurance Policy under the Dwelling Policy Form covers only the *fair value* or the repair costs, whichever represents the lower amount. It is thus not a replacement value insurance. In addition, the *insurance sum* is limited to USD 250,000.00 within the SFHA and to USD 500,000.00 outside of the SFHA[190]. Thus, there is an *insurance or liability limit.*
- The premium depends on the insurance sum, type of building, age of building, location of building and duration of ownership. In the SFHA, the height of the building is also a criterion[191].
- The excess amounts to between USD 500.00 and USD 1000.00, as the case may be[192].

The NFIP is affected by *considerable problems* of *charity* and *moral hazard* as well as by *negative selection.* The cause for this is, among other things, the *lack of solidarity.* This aspect will be discussed on p. 74 et seq. below.

[189] Consorcio, Diversity of Systems, p. 185.
[190] Consorcio, Diversity of Systems, p. 185 et seq., 191.
[191] Consorcio, Diversity of Systems, p. 190.
[192] Consorcio, Diversity of Systems, p. 191.

B. Florida Hurricane Catastrophe Fund (FHCF)

The FHCF was also founded after private insurance companies had withdrawn from the market for storm risks due to a natural disaster, namely Hurricane Andrew in 1992, and they cancelled numerous insurance contracts[193]. There is no institution for the protection against storm risks at federal level, but merely at the level of the individual states at best (another example is the Hawaii Hurricane Relief Fund).

The FHCF is a tax-exempt state fund managed by the State Board of Administration of Florida ("SBA"). It takes on the function of a reinsurance. Direct insurance is still underwritten by private insurance companies. All private building insurers operating in Florida *must* join the reinsurance program of the FHCF. They assign a portion of their risk, according to fixed criteria, to the FHCF within the framework of an excess of loss reinsurance and pay a premium for this[194].

Insurance applies to the *natural hazard of storms,* insofar as the storm is *classified as a hurricane* by the National Hurricane Center[195].

If the FHCF funds are not sufficient to meet its obligations (however, private insurers pay the first part of the loss in the amount of their retention), the SBA can issue debentures[196].

Here, too, there is *no compulsory insurance whatsoever* for building owners, let alone statutory insurance. Furthermore, building owners who are willing to take out insurance can be rejected by a private insurance. It is precisely those buildings that are located in risk zones which often fail to obtain insurance on the private market. Therefore, the *state-owned Citizen Property Insurance Corporation* ("Citizen") was created as a *lender of last resort,* which is supposed to absorb the residual market. Citizen is the best "customer" of the FHCF.

Citizen is financed by the insurance premiums of the contracts underwritten by it and a surcharge on all building insurance agreements by private insurance companies, even if the building is located outside a risk zone[197].

Even the FHCF and Citizen are affected by *considerable problems* of *charity* and *moral hazard* as well as *negative selection.*

---

193 KUNREUTHER/MICHEL-KERJAN, in: RICHARDSON/GORDON/MOORE II, p. 16 et seq.; CONSORCIO, Diversity of Systems, p. 198.

194 CONSORCIO, Diversity of Systems, p. 198.

195 CONSORCIO, Diversity of Systems, p. 199.

196 CONSORCIO, Diversity of Systems, p. 199.

197 KUNREUTHER/MICHEL-KERJAN, in: RICHARDSON/GORDON/MOORE II, p. 22; CONSORCIO, Diversity of Systems, p. 199.

2. Hurricane Katrina and the consequences

A. Basic problems of charity and moral hazard and negative selection

Since there is no compulsory insurance in the USA with respect to the natural hazards of flooding and storm, the insurance density is very low. This leads to the fact that ad hoc state aid has to be provided to a considerable extent in case of a natural disaster. It is precisely the prospect of state aid, however, that creates the wrong incentives in the sense of *charity hazard,* due to the fact that building owners do not attempt to obtain insurance protection, instead relying on state aid.

The same circumstance also results in the fact that building owners *are not motivated* to take *preventive measures.* There is no obligatory insurance solution which compels building owners to take preventive measures, whereby failure to do so is sanctioned by exclusion from coverage or whereby prevention is rewarded by lower retentions. Destroyed buildings are, if there are enough funds, rebuilt in the same place, even in risk zones. The repeated destruction of the same building is not a rare event[198].

The NFIP or the FHCF and Citizen are, in addition, affected by considerable problems of *negative selection.* Building owners *outside of* risk zones often do not take out insurance or, with respect to the storm risk, not from Citizen. Insurance is therefore taken out mostly by bad risks. This therefore results in an accumulation of bad risks with the NFIP as well as with Citizen, with the insurance density being too low even for bad risks.

Negative selection has led to, and continues to lead to, severe financial problems. In the last ten years before Katrina (prior to 2005), FEMA had to accept loans from the Treasury four times in order to support the NFIF just to keep it solvent. As a consequence of Katrina, the loans had to be considerably extended again. The NFIP is therefore sitting on a pile of debts amounting to about USD 17 billion[199].

At the end of 2004, the FHCF and Citizen were almost insolvent following a hurricane season with heavy losses. The FHCF has proved to be unsuitable for natural disasters as it had already reached its limit of performance, even though it had only absorbed 10% of the entire losses of USD 22 billion. In late 2005, following another severe hurricane season, a deficit in the amount of USD 1.6 billion had ultimately

---

198 KUNREUTHER/MICHEL-KERJAN, in: RICHARDSON/GORDON/MOORE II, p. 18 et seq., 21 et seq. see the reason for the lack of incentives for damage prevention in the subsidised premiums and propose risk-adjusted premiums where prevention measures are rewarded by premium rebates. Within the framework of the NFIP, rebuilding in risk zones is even encouraged by the Increased Cost of Compliance Mechanism, since affected building owners receive additional financial support for rebuilding; cf. CONSORCIO, Diversity of Systems, p. 191.

199 CONSORCIO, Diversity of Systems, p. 190 et seq.

accumulated which is now being reduced through a variety of measures (in particular debentures)[200]. Citizen's deficit amounted to USD 515 million at the end of 2004 and had to be compensated by additional surcharges on all insurance agreements. Citizen's insurance premiums are considerably reduced and are therefore unable to build up sufficient reserves for loss events. As a consequence of Katrina, Citizen's deficit grew again by USD 1.7 billion and is now being reduced through a variety of measures[201]. In addition, the FHCF had granted Citizen extremely favourable reinsurance terms for 2006, a fact that was criticised, however, in view of the strained financial situation of the FHCF itself[202].

The USA's protection systems regarding the natural hazards of flooding and storms are suffering from considerable "design defects" and are therefore *unable to maintain the financial equilibrium.* The severe gaps and deficits must, where the national budget permits this at all, be redressed by general taxpayer funds and ad hoc state aid in the amount of billions of dollars. The cause of the failure lies in *lack of solidarity.*

B. Lack of solidarity in the US system and what can be done about it

There is no balancing of good and bad risks, either within the framework of the NFIP or the FHCF. With regard to the storm risk, an accumulation of bad risks at Citizen is even promoted. As a result, both systems lose their financial equilibrium in loss events.

In addition, no equalisation takes place with regard to insured risks. There is no comprehensive insurance against natural hazards such as the KGV which covers all types of natural disasters and thus allows for risk diversification (all-risk coverage). Kunreuther/Michel-Kerjan rightly argue the case for an all-risk coverage in their analysis of Hurricane Katrina[203]. Likewise, the considerations regarding the introduction of compulsory insurance are heading in the right direction. Kunreuther/Michel-Kerjan state in this regard: *"Mandatory coverage would address many difficult problems. Since all residents would be financially protected, it would reduce the demand for government aid."*[204].

If the objective is comprehensive, extensive insurance protection accessible to everyone at affordable premium rates—and due to climate change, this goal will have to be pursued—then both the NFIP and the FHCF will have to introduce the *principle of solidarity.* The following should be done in order to ensure maximum solidarity:

---

200 Kunreuther/Michel-Kerjan, in: Richardson/Gordon/Moore II, p. 16.
201 Kunreuther/Michel-Kerjan: in: Richardson/Gordon/Moore II, p. 22 et seq.
202 Kunreuther/Michel-Kerjan: in: Richardson/Gordon/Moore II, p. 23.
203 Kunreuther/Michel-Kerjan, in: Richardson/Gordon/Moore II, p. 24 et seq., 30.
204 Kunreuther/Michel-Kerjan, in: Richardson/Gordon/Moore II, p. 29.

- *compulsory insurance* for all building owners should be introduced. Compulsory insurance must apply directly to the building owner, regardless of whether the community has joined the NFIP, the building is located in a risk zone, or the community is operating "floodplain management".
- the risk carrier, e.g. Citizen, must be built up to become the *statutory insurance* for the entire national territory. Accordingly, Citizen should be granted a legal monopoly.
- in general, the "floodplain management" and the prevention of damage as the result of natural hazards should be actively promoted by developing the insurance terms (exclusion from coverage, excess, premium rebate).

Only in this way will it be possible, particularly in view of climate change, to ensure comprehensive and affordable insurance protection for all building owners in a financially self-supporting manner. This is the prerequisite for rebuilding within a useful period of time, for strengthening the national economy, and for avoiding state aid.

C. The consequences of the "design defects" using the example of Hurricane Katrina

The effects on New Orleans, the largest city affected in the US state of Louisiana, are described below. Katrina, a category 5 hurricane, struck the Louisiana coast on 29 August 2005 at wind speeds of over 172 miles/hour. The eye of the storm came within 23 miles of the city centre of New Orleans. Due to the hurricane, *widespread flooding* [205] occurred in New Orleans. With respect to insurance, the *NFIP* was therefore called in.

The immediate *starting situation* after the catastrophe emerged as follows[206]:

- Numerous building owners in New Orleans did not have any insurance protection against flooding.
- Even in the risk zones (SFHA), a total of less than 50% of all buildings were insured.
- Of these, about 50–60% was subject to compulsory insurance due to specific mortgage loans. The compulsory insurance, however, was only complied with at a level of about 75% which, based on 100%, leads to a mere insurance density of about 22%.

---

205 Bostic/Molaison, in: Richardson/Gordon/Moore II, p. 253.

206 Kunreuther/Michel-Kerjan, in: Richardson/Gordon/Moore II, p. 20 et seq.

- Of the remaining 40% (approx.) of buildings in the SFHA that were not subject to compulsory insurance, only about 20% had insurance protection which, based on 100%, leads to an insurance density of only 4%.

This circumstance has had the following effect on the *rebuilding* and the *population* of New Orleans:

In 2007, large parts of the city were still uninhabited and abandoned, and countless houses had been given up in a state of destruction. Even as late as 2009, the condition of certain boroughs is described as follows: *"The tour through the district is nonetheless depressing, for the new buildings are oases in the midst of a dismal landscape of formerly developed and now vacant properties as well as disintegrated houses that are overgrown by plants (translation, original quote in German)."*[207].

- Prior to Katrina, New Orleans had a population of about 500,000 people. According to an estimate by the Louisiana Department of Health and Hospital, the population in 2006 was only about 210,000 people, that is, less *than half compared to the situation before Katrina. An investigation based on changed mailing addresses showed that in 2006 about 160,000 households had changed their address, with only 17,000 having a new address in New Orleans. However, about two thirds, that is, about 1*05,000 households, reported a new address in a different US state[208]. In 2009, New Orleans had a population of about 300,000[209]. This development leads to the conclusion that about 40% of the original population have left New Orleans permanently.
- The fact that even by 2009, that is to say, four years after Katrina, large parts of New Orleans have not yet been rebuilt shows the NFIP's ineptitude in dealing with a disaster situation. One of the critical requirements that a protection system must meet is to ensure rebuilding within a relatively short period of time.
- In view of this situation, the complete rebuilding of New Orleans is not a matter of course. Instead, the discussion in the USA today revolves around whether rebuilding should take place and what should be rebuilt[210]. With respect to the strategic starting situation, Bostic/Molaison state for example: *"From an economic perspective, given the tremendous economic disruption, it would seem unlikely that the post-disaster level of activity will reach pre-storm levels."*[211].
- Currently (autumn 2009), numerous efforts by the city of New Orleans along with FEMA and a wide variety of charitable foundations and volunteer organi-

---

207 Buenos dias, Nuevo Orleans, Basler Zeitung online, 23 October 2009. LOGAN, in: RICHARDSON/GORDON/MOORE II, p. 285.

208 LOGAN, in: RICHARDSON/GORDON/MOORE II, p. 279, 282 et seq.

209 Buenos dias, Nuevo Orleans, Basler Zeitung online, 23 October 2009.

210 Cf. BOSTIC/MOLAISON, in: RICHARDSON/GORDON/MOORE II, p. 265 et seq.: "What should be rebuilt?"

211 BOSTIC/MOLAISON, in: RICHARDSON/GORDON/MOORE II, p. 269.

sations are under way to advance rebuilding[212]. The fact that, four years after Katrina, New Orleans is still far removed from normality is shown by a statement made by the DHS superintendent, Janet Napolitano, made in August 2009: *"Our commitment to the Gulf Coast remains unwavering and our determination to bring to completion many of the projects is still underway. My goal is to eliminate the red tape, help rebuild now and rebuild the region stronger than ever."* [213] The future will show whether New Orleans will ever recover from Hurricane Katrina.

In view of the permanent devastation and gashes left by Hurricane Katrina in New Orleans which, as of this day, have not yet been overcome due to the lack of an adequate protection system, it is not surprising that parallels are being drawn with a natural disaster in the Third World[214]. Katrina makes it obvious that comprehensive insurance against natural hazards, based on statutory insurance, represents a necessity today and all the more so in the future.

[212] To cope with the consequences of Katrina, FEMA has set up a special "Louisiana Transitional Recovery Office". Cf. the numerous aid programmes at www.neworleans.com/community/rebuilding, www.fema.gov/hazard/hurricane/2005katrina and www.louisianarebuilds.info (as of 23 October 2009).

[213] U.S. Department of Homeland Security, FEMA, "2009 Louisiana Katrina/Rita Recovery", August 2009, available at www.fema.gov/hazard/hurricane/2005katrina, heading "Rebuilding Communities, Reuniting Families" (as of 20 October 2009).

[214] Clay, in: Richardson/Gordon/Moore II, p. 230 et seq.: "Katrina: a Third World catastrophe?"

II. The appropriate response to climate change

Based on this starting situation, the following interim conclusion can be drawn:

1. Type 3: State aid and optional insurance according to market supply

For the reasons shown above, ad hoc state aid as a protection system is far from sufficient. In no way does it meet the requirements.

In addition, there is consistent market failure for insurance against natural hazards. The market is unable to ensure a supply that meets the requirements. If insurance against natural hazards is left to the interplay of supply and demand, this inevitably results, for some of the buildings, in exclusions from insurance or coverage, unaffordable premiums and/or a system that is financially unstable. Government or state absorption institutions or reinsurance solutions that are not flanked by an compulsory insurance or, even better, are organised as statutory insurance, quickly lose their financial equilibrium and offer no protection at all in a disaster situation because the insurance density is much too low. As was shown above using the example of the USA, such systems are altogether useless.

Type 3 is therefore not an appropriate protection system when faced with climate change.

2. Type 2: Compulsory insurance

The compulsory insurance system has certain weak points:

- Comprehensive insurance protection can be undermined if coverage depends on a basic agreement for which itself no obligation exists.
- Negative selection can occur, whereby the financial equilibrium of the insurance organisation and thus the financial feasibility of the insurance system are not ensured.
- The incentives for the prevention of damage due to natural hazards, particularly on the part of the insurance organisation, are too little or do not even exist at all.
- Compulsory insurance alone guarantees neither comprehensive, complete and across-the-board insurance protection nor affordable premiums for all risks. This can, however, be improved by appropriate corrective regulations that create solidarity.

Type 2 does not quite meet the requirements even though certain corrections can be made by means of regulations.

3. Type 1: Statutory insurance

The statutory insurance system meets the requirements for a protection system:

- Statutory insurance automatically leads to comprehensive solidarity among policyholders and, in the case of the Swiss example, the KGV, on the part of the insurance organisations as well.
- The main objective of comprehensive, complete and across-the-board insurance against natural hazards can thus be achieved.
- Affordable premiums for all can thereby be ensured as well.
- Negative selection is automatically prevented since all risks (good and bad) are inevitably pooled in the same statutory insurance. The financial stability and financial feasibility of the system and of coverage is thereby ensured.
- Strong incentives are created, both on the part of the insurance organisation as well as on the part of the policyholders, to pursue the prevention of damage caused by natural hazards and to invest in damage abatement (insurance organisation). Both sides benefit from this; the system as a whole is sustainably strengthened. The problem of moral hazard is prevented.
- The elimination of the market or of competition, however, does not constitute a disadvantage. As stated above, market failure has been identified. Even with the compulsory insurance system, intervention is required by means of regulations to a degree that excludes the market in order to make the system functional. Product competition is not possible and is not necessary either, for the product is already defined based on the purpose of comprehensive insurance protection. The only thing that matters is that protection against damage due to natural hazards, which is in the public interest, is guaranteed as efficiently as possible. This can be optimally ensured with the statutory insurance system. Unnecessary costs for sales and marketing are saved.

Type 1, and therefore, the Swiss KGV system regulated by public law, as well as the Spanish Consorcio system, meet the requirements excellently when faced with climate change. The insurance system is further strengthened by the integration of the prevention of damage due to natural hazards and damage abatement into the insurance organisation. When faced with climate change, a system according to type 1 is simply a necessity.

Chapter 4: Legitimacy of statutory insurance against natural hazards using the example of the KGV

I. Admissibility under Swiss law

1. Starting situation—cantonal legal monopoly

As stated above, a statutory insurance system requires obligatory insurance (compulsory insurance); the monopoly of the insurance organisation, in this case the KGV, is characteristic of this system. The obligatory insurance in itself is compatible both with the constitutional guarantee of the right of ownership (art. 26 BV) as well as with economic freedom (art. 27 and 94 BV)[215]. The legitimacy does not give rise to discussions which is why this point will not be discussed further. Below, we will focus our attention exclusively on the legitimacy of the KGV monopoly in insurance of buildings against natural hazards. The latter is anchored in the respective cantonal GVG, e.g. with the following wording:

GVG ZH[216]*:*

"§ 10 Obligatory insurance

All buildings within the canton must be insured with the institution.

§ 14 Prohibition of double insurance

The buildings, or property that is similar to buildings, that are insured with the institution must not be insured elsewhere for the benefits provided for in this law (translation, original quote in German)."

GVG BE[217]*:*

"Art. 3 Obligatory insurance

The buildings located in the Canton of Berne must be insured with the building insurance for those hazards that may be covered thereby and must not be insured elsewhere for the same hazards (translation, original quote in German)."

The respective titles of the statutory provisions merely contain the term *"obligatory"*. The fact that, in addition to the obligatory insurance, a monopoly also exists is apparent from the prohibition of double insurance, or the prohibition to *"insure"* buildings *"elsewhere"*. Through this directive, private insurers are indirectly excluded from the market as an insurance offer does not make any sense since the owner of the building is not permitted to accept it (the "market", however, is of theoretical nature only for,

215 For details, see MÜLLER, Kommentar GV, p. 39 et seq.

216 Cf. footnote 151 above.

217 Building insurance law of the Canton of Berne as of 6 June 1971 (BSG 873.11).

without the KGV, the federal regulation, which likewise does not allow for a market but rather for a state market and price system, would apply; cf. p. 64 et seq. above). The directive is aimed at the building owner, not at other insurances. The monopoly is therefore qualified as an *indirect* monopoly (in contrast to a direct one)[218]. Furthermore, the monopoly is based on a legal proposition and is therefore referred to as a *legal* monopoly[219]. Finally, a *government (state)* monopoly exists since, in the present case, a public-law institution has the monopoly[220].

The KGV thus fulfil their mandate within the framework of a legal, indirect and government monopoly.

When examining legitimacy, compatibility with constitutional economic freedom and with the KG has priority.

2. Compatibility with economic freedom

A. Economic freedom as part of a socially and environmentally committed economic system

Economic freedom is incorporated in art. 27 and 94 BV. The provisions are as follows:

Art. 27 Economic freedom

1 Economic freedom is guaranteed.

2 It encompasses in particular the free choice of profession as well free access to gainful private employment and the right to freely carry out such occupation.

Art. 94 Principles of the economic system

1 The Confederation and the Cantons abide by the principle of economic freedom.

2 They safeguard the interests of the Swiss economy as a whole and, together with the private sector, contribute to the welfare and the economic security of the population.

3 Within the framework of their responsibilities, they ensure advantageous framework conditions for the private sector.

4 Deviations from the principle of economic freedom, particularly including measures that are directed against competition, are permissible only if they are provided

---

218 MÜLLER, Kommentar GV, p. 43 et seq.; HÄFELIN/MÜLLER/UHLMANN, p. 287, 550.

219 This is in contrast to a de facto monopoly which is based on de facto circumstances, e.g. the state's sovereignty over the public cause, which grants the state a supply monopoly for grid-bound energy supply depending on the arrangement of the energy sector. MÜLLER, Kommentar GV, p. 43; HÄFELIN/MÜLLER/UHLMANN, p. 287, 550.

220 Cf. MÜLLER, Kommentar GV, p. 43.

for in the federal constitution or if they are based on the prerogative rights of the cantons.

(translation, original quote in German)

A controversy is being argued in the doctrine as to whether a policy decision in favour of a market-based and competition-oriented economic system was made by means of these provisions. Furthermore, those who affirm this have diverging views on specifically what this means or what consequences this has.

VALLENDER takes the view that the BV contains a basic regulatory decision in favour of a market-based and competition-oriented private sector. He specifies, however, that the decision in favour of a specific system based on a theoretical model, or an economic system, is not associated with this[221]. In addition, he explains that, although competition as a concept has a special significance, it does not comprise a blanket order to protect competition and no absolutisation of competition[222]. According to VALLENDER, the establishment of new cantonal monopolies is still possible, even if these are not prerogative rights. A precondition is that the requirements of art. 36 BV are met and that no fiscal purpose (pursuit of profit for the benefit of the treasury) is associated with the monopoly[223]. VALLENDER/HETTICH/LEHNE, on the other hand, hold the opinion that, in view of the (alleged) basic regulatory decision, socio-politically motivated monopolies are questionable and that monopolisation would be a possibility only if it was mandatory for a legally protected interest of supreme importance. Thus, the latitude for government monopolies is thus given a narrow range here on the grounds of basic considerations[224] .

The views of RHINOW and RICHLI are convincing. Both demonstrate that economic freedom, or the economic system, is embedded in the overall constitution and that it is characterised by a multitude of social and environmental obligations. Within the BV, there is neither a self-contained economic constitution nor does it contain an express commitment to a specific economic system or economic model. The BV thus does not contain any basic regulatory decision in favour of a market-based and competition-oriented private sector. Instead, according to RHINOW, it contains a clear commitment to social obligations, a fact that is already apparent from the preamble[225] and the purpose clause of the BV (art. 2)[226], as well as from art. 94 itself,

221 VALLENDER, St. Galler Kommentar BV, BV 27 N 85 et seq., BV 94 N 14 et seq., 25.

222 VALLENDER, St. Galler Kommentar BV, BV 94 N 20.

223 VALLENDER, St. Galler Kommentar BV, BV 27 N 72 et seq.

224 VALLENDER/HETTICH/LEHNE, p. 186.

225 This is apparent from the wording "*the strength of a people is measured by the well-being of the weak*". The environmental obligation arises out of "*in the responsibility for creation*". (*Translation, original quotes in German)*

226 This is apparent from various formulations of the purpose clause, e.g. "*It [Switzerland] promotes the common welfare, the sustainable development...*" "*It advocates the lasting conservation of natural*

namely art. 94 par. 2 BV[227]. Thus, the economic constitution does not merely consist of economic freedom but also comprises numerous other constitutional standards regarding environmental protection (art. 73 et seq.), public works (art. 81 et seq.), social security systems (art. 111 et seq.), business cycle policy (art. 100 et seq.) and social goals (art. 41)[228]. In this context, RICHLI refers to a *socially and environmentally committed market economy* as well as a *liberal, competition-oriented, socially and environmentally committed constitutional state with a corresponding economic system*[229]. Therefore, a hierarchy between the fundamental economic rights on the one hand and their restriction by the government on the other hand does not exist, and this complies with the constituent's intent[230]. Overall, one has to concur with RHINOW, who explains[231]:

"When interpreting the principle of economic freedom and outlining the Swiss economic system, the whole constitution has to be kept in sight. Economic freedom and the arrangement of the private sector are not a purpose to themselves; on the one hand, they answer to the basic needs of a liberal and self-responsible way of life and, on the other hand, they also represent a means of achieving fair living conditions and of promoting common welfare as well (translation, original quote in German)."

This is also the view of the Federal Court when it states in its decision on the monopoly of the KGV GL[232]:

"The market-based system, which is protected by the institutional or economic-political competency of art. 31 BV, finds its legitimation in the fact that the economic necessities of the population are thereby satisfied as efficiently and as affordably as possible (translation, original quote in German)."

Competition and economic freedom do not serve an end in itself but instead are a means to promote the well-being of the population[233].

resources..." (translation, original quote in German).

227 Based on this provision, the Confederation and the Cantons, in addition to the private sector, should contribute to the welfare and the economic security of the population. RHINOW, Verfassungsrecht der Schweiz, p. 569.

228 RHINOW, Verfassungsrecht der Schweiz, p. 565 et seq. Likewise, BIAGGINI/LIENHARD/RICHLI/UHLMANN, Wirtschaftsverwaltungsrecht, p. 10.

229 RICHLI, Wirtschaftsverfassungsrecht, p. 15 et seq.

230 RICHLI, Wirtschaftsverfassungsrecht, p. 16.

231 RHINOW, Verfassungsrecht der Schweiz, p. 572.

232 FCR 124 I 25, 31.

233 In this context, it is appropriate to refer to the discussion under the heading "more economic approach". It takes place within the rules on competition when the assessment of restraints on private competition pursuant to the antitrust law is involved (competition agreements, market domination and company mergers). This is a different level from the one discussed here as, in the present case, the issue is the embedding of the principle of competition in the overall constitution and its significance with respect to other constitutional standards. Furthermore, for the compulsory insurance and statutory insurance systems existing in Switzerland, a statutory market and pricing system prevails that

One would have to agree with the opinion of RHINOW and RICHLI. It is apparent from the explanations above that the BV does not contain any basic regulatory decision in favour of a market-based and competition-oriented private sector from which legal

---

excludes this system from the scope of the KG from the start (see p. 67 et seq. and 96 et seq.). It is, however, easy to imagine that the discussion also "spills over" to the current level. In the EU, the "more economic approach" is increasingly gaining ground in the application of competition law, e.g. in the assessment of whether the misuse of a market-dominating position pursuant to art. 102 TFEU [82] exists. The "more economic approach" evaluates restraints on competition from the point of view of economic results. Whether a restraint on competition is proscribed should be determined based on empirical economic data, analyses and results. If these indicate an increase in prosperity in spite of restraints on competition, or if they indicate that the omission of the restraint on competition results in a decrease in prosperity, then the restraint on competition is permissible. This is based directly on the economic result. The proponents of this approach claim that they are thereby applying an objective method. The traditional concept of competition law, by contrast, evaluates restrictions only from the aspect of freedom of action and freedom of competition of the individual. If this is restricted, there will be a proscribed restraint on competition. This concept is based on the premise that competition will automatically lead to an increase in prosperity. Hence, the actual economic result can and should not be an issue. Competition should be protected as an individual right and an institution. Ordoliberalism is the source of the traditional concept. It is based on an axiom, a premise introduced without proof that competition will always lead to more prosperity. The specific rules for the evaluation of restraints on competition are derived from this. The proponents of the "more economic approach", however, take empirical observations as their starting point, derive general rules from this and construct corresponding hypotheses. The objective is to examine whether the effects on prosperity, particularly efficiency advantages, are effectively negative or are instead positive. However, the proponents of the traditional concept argue that the question of what represents prosperity and an increase in prosperity ultimately cannot be proven objectively but that premises are also assumed in this regard. In addition, efficiency could be determined not only based on cost alone but other criteria such as quality would have to be considered as well. Even the "more economic approach" would have to resort to value judgments. Furthermore, the required economic analysis would raise unsolvable practical problems for public authorities and courts that apply the law. Under the traditional concept, competition is protected for its own sake. The "more economic approach", however, regards competition only as a means for the purpose of increasing prosperity, with the consequence that if competition does not result in an increase in prosperity or even leads to a decrease in prosperity, open competition will have to take a back seat. Even though this discussion does not relate to the current level, it still allows lessons to be drawn from it that can be made productive for a public-law monopoly. Even with regard to the issue of whether such a monopoly is justified, it definitely makes sense, based on economic analyses and results (having regard to qualitative aspects), to examine whether an increase in prosperity or, to the contrary, a reduction in prosperity results if *no* public-law monopoly exists but if open competition prevails instead. Based on the numerous instances of privatisation and deregulation in recent years in Switzerland and abroad, empirical data that allow statements to be made are available or can be collected. Should the experience gathered yield the finding that competition does not result in an increase in prosperity, which refers to an improvement with regard to qualitative aspects (more specifically, quality, service, accessibility, continuity of service, that is to say, not just costs), then a public-law monopoly which is better able to guarantee prosperity, is given preference. Open competition as such does not automatically guarantee more prosperity, aside from the fact that actual competition does not even occur in situations of market failure, but that merely private monopolies are promoted. The incorporation of an economic analysis thus definitely makes sense, even though the economic perspective must not be absolutised; that would be just as misguided as the absolutisation of competition as an end in itself. KÜNZLER has dealt with the controversy between "more economic approach" and traditional concept extensively. A concise outline can be found in ZÄCH/KÜNZLER, EuZ 2009.

limitations for government monopolies could be derived. The existence or the establishment of state-controlled cantonal monopolies is therefore *not* restricted or made impossible by the basic orientation of the BV. This also complies with the case law of the Federal Court (see p. 89 et seq. below).

B. No right to protection from government competition or subsidiarity of government economic activity

Moreover, no claim to protection from government competition can be derived from economic freedom. The government is allowed to pursue economic activities[234]. In addition, no subsidiarity principle exists in the BV to say that a task should be performed by the government only if the same task cannot be performed equally well by the private sector[235]. Instead, it is solely up to the legislature or the political decision-making process whether a task should be performed as a public task by the government or by the private sector. This was explicitly decided thus by the Federal Court[236]:

"Most government activities can in principle be performed by the private sector as well. Which tasks should be considered government or public tasks is largely a political question and subject to the change in opinions, and is therefore primarily the responsibility of the legislature [...]. Although economic freedom contains an institutional component whereby economic activity should be regulated on the basis of competition (art. 94 par. 4 BV), it does not follow from this that the government is prohibited from declaring a certain task, which per se could also be attended to by the private sector, to be a public one (translation, original quote in German)."

234 BIAGGINI, Verfassungsrecht der Schweiz, p. 785; follows from VALLENDER, St. Galler Kommentar BV, BV 94 N 6; VOGEL, Der Staat als Marktteilnehmer, p. 102 et seq.; UHLMANN, p. 180 et seq.

235 Contrary to LIENHARD, in: COTTIER/OESCH, p. 394 et seq., such subsidiarity cannot be derived from economic freedom and much less so from art. 5a BV. Art. 5a and the associated art. 42, 43 and 43a BV relate exclusively to the subsidiarity of the performance of a task by the Confederation as opposed to the performance of a task by the cantons. According to communication NFA, these provisions expressly do not refer to the government—private sector relationship ("Within the framework of the current reform of federalism, the principle of subsidiarity confines itself to the relationship between the levels of government. Any extension to the government—private relationship would inevitably have to raise a wide range of fundamental sociopolitical issues and thus go beyond the scope of the financial equalisation reform *(translation, original quote in German)*." Communication NFA, BBl 2002, 2291, 2457. Art 5a and 43a BV were introduced within the framework of the reorganisation of national financial equalisation and are intended to secure the federalist state structure and make a central state impossible. For art. 5a and 43a, cf. SCHWEIZER/MÜLLER, St. Galler Kommentar BV, BV 5a N 9 et seq. as well as SCHWEIZER/MÜLLER, St. Galler Kommentar BV, BV 43a N 7.

236 FCR 2P.67/2004 E. 1.6. As a result, likewise VALLENDER/HETTICH/LEHNE, p. 192 who, although they advocate the subsidiarity of government economic activity, ultimately state that the decision about that is the responsibility of the legislature.

Therefore, no legal claims, such as protection from government competition or subsidiarity of government economic activity, exist that would in principle be an obstacle to a government monopoly.

C. Statutory insurance as a monopoly that is justified in socio-political and police terms

Now that it has been resolved that the federal constitution and the economic constitution are not opposed to a government monopoly in principle, it remains to be examined whether the KGV statutory insurance meets the legal requirements for a government monopoly.

a) Legal requirements for compatibility with economic freedom

A government monopoly is compatible with economic freedom if it is *compliant with the principle* and if it meets specific additional requirements. A monopoly that is based on a restriction of economic freedom purely on economic-political, professional-political or structural-political grounds is *not compliant with the principle* (corresponding measures are also termed *economic-political* measures). This refers to interventions that *exclusively* serve to favour certain providers or to maintain a certain economic structure. Such monopolies would be permissible in accordance with art. 94 par. 4 BV only by virtue of a basis in the BV or based on a prerogative right of the cantons[237].

However, measures such as government, particularly cantonal, monopolies are compliant with the principle if they can invoke *public interest.* For example, the pursuit of *non-profit, charitable, socio-political or police purposes* is recognised as being of public interest, for example. Monopolies stipulated on the basis of such purposes do not deviate from economic freedom and therefore do not come under art. 94 par. 4 BV[238]. They are not opposed to economic freedom. In this context, BIAGGINI refers to *police and welfare monopolies*[239]. The judicature has commented on this as follows[240]:

> *"Non-profit and charitable efforts take effect where economic competition leads to undesirable socio-political results. They do not follow the market and competition rules from the outset, which are protected by art. 94 BV. In the pursuit of non-profit and charitable purposes, the government's constitutionally-permitted leeway is greater than in the sector of economic-politically motivated interventions in commercial*

---

237 VALLENDER, St. Galler Kommentar BV, BV 27 N 49 et seq., BV 94 N 5 et seq.

238 VALLENDER, St. Galler Kommentar BV, BV 27 N 50, 72 and BV 94 N 5.

239 BIAGGINI, Verfassungsrecht der Schweiz, p. 786.

240 Judgment by the Administrative Tribunal of the Canton of Berne dated 11 July 2001 (VGE 21127) (BE) *with numerous references to the case law of the Federal Court.* BVR 2002 p. 123 et seq., p. 132.

competition. This also complies with the established case law of the Federal Court and the tenet whereby socio-politically motivated restrictions on economic freedom are not considered an inadmissible deviation (FCR 125 I 417 E. 4a p. 422, 123 I 201 E. 4 p. 205, 121 I 129 E. 3b p. 132, 120 Ia 126 E.4 p. 132 et seq.; BBl 1997 I 175, 291; GYGI/RICHLI, Wirtschaftsverfassungsrecht, 2nd ed. 1997, p. 98 et seq.; RENÉ RHINOW, *loc. cit., p. 313 et seq.). While the government should in principle refrain from intervention in the domain of economic competition, it is basically at the government's discretion which non-profit or charitable purposes it wants to serve or promote. Police or socio-political monopolies thus do not represent an inadmissible deviation from economic freedom pursuant to art. 94 BV (FCR 125 II 508 E. 5b p. 511, 124 I 11 E. 3b and 4a p. 15 et seq.; 124 I 25 E. 2 p. 27, 123 II 359 E. 5b p. 368) (translation, original quote in German)."*

The primary prerequisite for the compatibility of a government, cantonal monopoly is therefore that such a monopoly can invoke *public interest.*

The following additional conditions must be met for such a government cantonal monopoly serving the public interest to be permissible[241]:

- the monopoly must have a *sufficient legal foundation.*
- the monopoly must be *proportionate.*

b) Socio-political and police reasons and hence public interest exist

The Federal Court has commented on the monopoly of the KGV and on statutory insurance in a landmark decision. The crucial parts of the grounds of the judgment are rendered below[242]:

"What proportion of the total revenues of the insurance is used to cover losses is not the decisive factor insofar as the building insurances also finance preventive fire protection and thereby contribute to preventing the damage from happening, which is of benefit to the insured as well. In this sense, a correlation between the insurance system and the claims history is conceivable in any event: it appears to be quite unlikely that the claims history, which is significantly worse in cantons without monopoly insurance, should be purely coincidental. In contrast, it appears rather more plausible that the cantonal building insurances contribute to the significantly more favourable claims history by expending more on prevention, thereby making the low premium possible. [...]

241 The prerequisites result from art. 36 BV, whereby the preservation of the essence is not required. Cf. FCR 124 I 25, 27; MÜLLER, Kommentar GV, p. 45; VALLENDER, St. Galler Kommentar BV, BV 27 N 72.

242 FCR 124 I 25, 30 et seq.

It is also conceivable that certain administrative synergies with government tasks that must be performed anyhow are possible by means of monopolisation, and that certain administrative costs, namely acquisition costs, can be kept lower.

On the whole, there are plausible reasons for the assumption that the cantonal building insurances facilitate a significantly more favourable claims history, thereby making both lower losses as well as significantly lower insurance premiums possible.

For this reason, a sufficient public interest in monopolisation of building insurance can be considered to be established. The market-based organisation which is protected by the institutional or economic-political component of art. 31 BV finds its legitimation in the fact that it is intended to satisfy the economic needs of the population as efficiently and affordably as possible (ETIENNE GRISEL, Liberté du commerce et de l'industrie, vol. 1, Bern 1993, p. 86 et seq.; JÖRG PAUL MÜLLER, Die Grundrechte der schweizerischen Bundesverfassung, Bern 1991, p. 354, KLAUS *Vallender, Wirtschaftsfreiheit und begrenzte Staatsverantwortung, 3rd ed. Bern 1995, p. 85). If it turns out that a service, the mandatory nature of which—as the appellant herself recognises – invokes considerable public interest, can be rendered significantly more advantageously by a monopoly system than by the private sector, then this can represent a permissible public interest in order to justify a restriction of the freedom of trade and economic freedom (FCR 101 Ia 124 E. 8b p. 128). [...]*

Premiums can indeed be more favourable from a private insurance in a given case, in particular for major customers such as the appellant. But it does not follow from this that only a system—as proposed by the appellant—that permits private insurance for good risks, leaving the bad risks to the cantonal building insurance alone would be permissible. Such a provision would, with considerable probability, render the previous advantages of the cantonal building insurance void. A certain social equalisation is the essence of most socio-politically motivated regulations. The fact that this results in higher payments for certain policyholders in a given case is admissible on constitutional grounds at any rate as long as this additional burden is not excessive, which, however, is not demonstrated in the appeal (translation, original quote in German)."

On the one hand, the Federal Court recognises the positive effects of the interaction between statutory insurance and damage prevention[243]. The "securing and insuring" system is declared to be the positive reason for the favourable claims history, the low premiums and thus the efficiency of the monopoly system. This results not only in

[243] It makes no difference that the FC only refers explicitly to fire prevention and not to prevention of losses caused by natural hazards. The positive interaction is the same. Furthermore, the FC mentions "damage prevention", which includes the prevention of damage caused by natural hazards, as a neutral term in several instances. The legal conclusions of the FC therefore apply to the connection between statutory insurance and damage or loss prevention in all areas.

social but also in police reasons that constitute public interest. Damage prevention and high-quality damage abatement protect persons and tangible assets, both of which are police goods.

In addition, the positive *synergy effects* between statutory insurance and damage prevention / damage abatement as well as the advantage that no sales and marketing costs (acquisition costs) are incurred are put in the balance.

Moreover, the *balancing of good and bad risks* (the *solidarity*) is acknowledged positively and is referred to as a feature that proves precisely that the monopoly of statutory insurance is *socio-politically justified.* The objective is to make comprehensive insurance possible for all building owners, regardless of risk, with affordable terms and premiums, and on the basis of a financially stable, self-supporting and efficient insurance system, and this also requires appropriate damage prevention and damage abatement. The appellants' proposal to provide for a system that reserves the good risks for the private insurance companies and leaves the bad risks to the state KGV was rejected by the Federal Court with good reason, since the advantages of the monopoly system would be destroyed by precisely that. It is exactly this "cherry-picking" that is not permissible. As can be inferred from the preamble, art. 2 and art. 94 BV, it is the state's responsibility to ensure the welfare of the (entire) population. Welfare is not achieved by the state absorbing the losses arising from bad risks while the private sector skims off the profits from good risks. It is instead achieved by a system that is self-supporting due to the equalisation of risks and that does not unilaterally burden the state with losses that will ultimately have to be borne by the population.

When the Federal Court concludes that the monopoly system safeguards comprehensive insurance protection in a *significantly more advantageous* manner, it thereby refers not only to the lower premiums (according to the FC, the KGV premiums are about 40% lower on average compared to private insurance premiums)[244], but also to the other significant advantages—more favourable claims history, better prevention and damage abatement and synergy effects. These advantages will gain even more in importance in the future.

Hence, socio-political and police reasons, and thus public interest, for the monopoly of statutory insurance are established[245].

The fact that comprehensive insurance coverage represents a public interest in itself is not even a subject of discussion by the FC but is simply assumed. What is considered recognised public interest, is subject to the change in opinions and values

244 FCR 124 I 25, 29 et seq.

245 Ditto MÜLLER, Kommentar GV, p. 47.

over time[246]. As the result of climate change, protection against the consequences will increasingly shift to the focus of public discussion. With the increase in natural disasters and losses caused by natural hazards, the need for comprehensive insurance protection for buildings will become ever more pressing. *Public interest in comprehensive insurance against natural hazards and thus in a system that ensures this will increase considerably due to climate change.* The accumulation of natural disasters in recent years (one of the most recent examples is the flood disaster in the Philippines in September/October 2009) shows that that there is already considerable public interest.

c) Sufficient legal basis

This premise requires the legal monopoly to be incorporated in a general, abstract and sufficiently specific standard (legal proposition requirement) as well as in legislation in the formal sense (legislation requirement)[247]. This requirement is met, as the GVG contain specific provisions in each case that unequivocally state the legal monopoly (see p. 83 et seq. above)[248].

d) Proportionality

The monopoly is proportional if three elements are fulfilled[249]:

- Suitability: the monopoly must be suitable for achieving the goal which is in the public interest.
- Necessity (need/prohibition of excessiveness): the monopoly must be required in order to achieve the goal which is in the public interest. If the goal can be achieved using an equally suitable milder measure, the latter measure must be applied.
- Reasonableness: the monopoly, or the associated exclusion of private insurances, must be reasonably proportional to the goal which is in the public interest.

The monopoly or the statutory insurance allows the goal of comprehensive, solidary, efficient insurance against natural hazards, which is accessible to all, to be achieved in a virtually ideal manner. *Suitability* therefore exists[250].

The necessity of the monopoly is disputed. VALLENDER holds that this element is not fulfilled as the *"public interest in obligatory insurance"* could *"also be achieved without nationalisation by the introduction of obligatory insurance (translation,*

---

246 HÄFELIN/MÜLLER/UHLMANN, p. 115, with reference to case law; TSCHANNEN/ZIMMERLI/MÜLLER, p. 149, with reference to case law.

247 Cf. HÄFELIN/MÜLLER/UHLMANN, p. 80 et seq., 83 et seq.

248 MÜLLER, Kommentar GV, p. 47.

249 HÄFELIN/MÜLLER/UHLMANN, p. 124 et seq.

250 As a result, likewise MÜLLER, Kommentar GV, p. 48 et seq.

original quote in German)"[251]. To begin with, this contradicts the above-mentioned landmark decision by the FC which states that obligatory insurance, partly combined with price supervision, was already currently in existence and did not ensure insurance that was as favourable as the monopoly system[252]. In addition, the FC states[253]:

"Apart from that, the fact that a different legal solution might also be conceivable does not render the solution chosen by the Glarn legislature disproportionate (FCR 101 Ia 124 E. 8b p. 129) (translation, original quote in German)."

In the first instance, the decisive issue is that public interest does not just amount to obligatory insurance . Public interest lies in achieving the above-mentioned goal. This cannot be achieved completely even by means of an obligatory insurance with price supervision since, not only are the premiums higher, but the claims history is less advantageous as well, which can be ascribed to the lack of integration of damage prevention / damage abatement, the lack of synergies and the costs of customer acquisition. Without a monopoly, for example, the actuarial measures could not be enforced, whereby the prevention of losses as the result of natural hazards would in turn be undermined. If the objective is to achieve comprehensive and efficient insurance against natural hazards, which is accessible to all (and this is intended by the legislature and represents a political decision to be respected), then a monopoly is required.

In addition, the above-mentioned goal reflects a public responsibility in accordance with the decision made by the legislature in the cantons that have KGV. Changing the system as suggested by VALLENDER, in other words reducing it to obligatory insurance that is implemented by private insurance companies, would thus mean privatisation. In view of the sobering results of certain privatisation and deregulation proposals (in particular in the energy supply sector[254]), utmost scepticism is

251 VALLENDER, St. Galler Kommentar BV, BV 27 N 79.

252 FCR 124 I 25, 31.

253 FCR 124 I 25, 31.

254 The implementation of the Energy Supply Act (StromVG, SL 734.7) and the Energy Supply Decree (StromVV, SL 734.71; as of 1 January 2008 and 1 April 2008) by which the electricity market is to be liberalised, immediately led to exorbitant price increases of more than 20% in some cases. This triggered a storm of complaints, in particular from industrial undertakings with high energy consumption, to the newly created ElCom regulator (Eidgenössische Elektrizitätskommission; Swiss Electricity Commission). By the end of 2008, the regulator had received more than 2,500 complaints and reclamations due to excessive electricity prices. In order to lessen the price increase, the Federal Council eventually took a variety of measures and amended the energy supply decree accordingly as of 1 January 2009. The perceived liberalisation has thus profoundly disappointed expectations regarding a lower electricity price from the start. In addition, there can be no talk of deregulation since, after all, numerous new, complicated provisions and a regulation and price monitoring authority (ElCom) had to be created. Cf. the press release of the ElCom dated 22 September and 5 December 2008 as well as 9 March 2009, available from www.elcom.admin.ch, as well as the countless press articles on the subject, e.g. "Hefty electricity price increases and many open questions *(translation, original quote in German)*" (NZZ Online dated 2 September 2008), "Federal Council believes that

therefore required with respect to the results of privatisation and deregulation. As accurately stated by UEBERSAX[255] years ago, privatisation leads to a loss of democratic control regarding a task of which the fulfilment continues to be in the public interest. In order to safeguard this control to some degree, a supervisory authority has to be created with great effort; thus, privatisation does not lead to deregulation but to an increase in regulation ("re-regulation"). Furthermore, a monopoly that had previously been clearly laid down by public law is usually replaced by a private monopoly or an oligopoly, as is the case in the cantons that do not have KGV. Here, the private insurers organised in the ES pool practically have a monopoly. There is also no competition as this has been completely eliminated by the statutory market and price system. Obligatory insurance without a market and price system, i.e. a situation where insurance coverage is left to the market forces leads to market failure. Bad risks would not find insurance protection, premiums would be unaffordable in some cases, and the goal of comprehensive insurance against natural hazards that is accessible to all would not be achieved. The quotation from the Federal Court judgment should indeed be interpreted in this sense. Another conceivable solution will not result in the disproportionality of a monopoly under public law, which in reality achieves the goal that is in the public interest in a very efficient manner.

Hence, the necessity exists[256].

With respect to *reasonableness,* it should be said that the goal stated above reflects a significant public interest. In view of this substantial public interest, which will continue to increase in importance, the state monopoly is a reasonable measure[257]. This holds even truer since the alternative would be the existing statutory market and price system in accordance with the federal regulation (see p. 67 et seq. above), which also does not allow for a competition-oriented private sector. As mentioned above, without any regulation or within the framework of the open market, the intended goal cannot be achieved at all due to market failure. The public interest described here should therefore have much greater weight attached to it than the merely theoretical, liberal activity of the private industry in this sector, so that a state monopoly to safeguard this interest is reasonable in any case.

The legal state monopoly of the KGV and hence the statutory insurance system is consequently proportional.

---

liberalisation is another reason for rising electricity prices *(translation, original quote in German)*" (NZZ Online dated 26 September 2008), "Electricity rates under high voltage *(translation, original quote in German)*" (NZZ Online dated 26 September 2008) as well as "Federal Council squeezes down on electricity rate increases *(translation, original quote in German)*" (NZZ Online dated 5 December 2008).

255 UEBERSAX, ZBl 2001, p. 402 et seq., 417 et seq.

256 As a result, likewise MÜLLER, Kommentar GV, p. 49 et seq.

257 As a result, likewise MÜLLER, Kommentar GV, p. 50 et seq.

e) Conclusion

The monopoly of the KGV or the statutory insurance system meets all legal requirements with respect to economic freedom. It is compatible with the economic freedom of the BV.

3. Compatibility with the antitrust law

Regarding compatibility with the antitrust law, we can primarily refer to p. 67 et seq. above. Even the statutory insurance system lays down the definitive economic parameters of a market. The scope of coverage, the premiums and all other terms are determined by the respective GVG. Pricing and product competition is thus not possible. Add to this the legal monopoly of the KGV which purposely excludes any competition. The GVG as cantonal[258] public-law legislation thus also constitutes a statutory market and price system in accordance with art. 3 par. 1(a) KG on the one hand (such as insurance against natural hazards provided by private insurers in accordance with the VAG), which deliberately excludes any competition and takes precedence over the KG. The KG does therefore not apply. The difference to the statutory market and price system of the private insurances lies in the fact that the KGV system is significantly more efficient.

On the other hand, the clause in art. 3 par. 1(b) KG is relevant with respect to the KGV, based on which regulations[259] are reserved, *"which provide individual companies with special rights for the purpose of performing public tasks (translation, original quote in German)."* Comprehensive building insurance against natural hazards, which is accessible to all, is a public task in the cantons that have KGV[260]. For example, a legal[261] monopoly can be considered a "special right", which was granted to the KGV precisely so they can perform their public task. In the present case, this excludes competition completely and does not leave any room for the KG to be applied[262]. For this reason as well, the KG does not apply.

The regulations that constitute the statutory insurance system of the KGV thus take precedence over the KG, with the consequence that the KG does not apply. In this sense, the statutory insurance system is compatible with the KG.

---

258 Even corresponding cantonal law takes priority over the KG. ZÄCH, Kartellrecht, p. 133 et seq.

259 These can be cantonal regulations as well. Carron, Commentaire Romand LCart, LCart 3 al. 1 N 22.

260 This is expressed on the one hand in the creation of the GVG and the statutory insurance system itself, and on the other hand by the fact that all KGV are tax-exempt since they pursue public purposes.

261 A de facto monopoly is not sufficient. ZÄCH, Kartellrecht, p. 136.

262 Cf. ZÄCH, Kartellrecht, p. 136 et seq.

4. Statutory insurance as a basic service provision within the meaning of art. 43a par. 4 BV?

Art. 43a par. 4 BV, which entered into force on 1 January 2008, reads as follows:

"Basic services must be open to all persons in a comparable manner (translation, original quote in German)."

According to the NFA communication, it is necessary that basic services be provided *"uniformly", "across the entire national territory"*[263]. In its report on the "Basic service provision in infrastructure (public service)", the SFC expressed what is meant by this[264]:

"According to the definition by the Federal Council, the public service encompasses a politically defined basic service provision with infrastructure goods and services that should be available to all levels of the population and to all regions of the country according to the same principles, of good quality and at reasonable prices.

The public service is thus equated with a basic service provision. According to the above definition, the basic service provision always includes basic facilities with infrastructure goods and services to be determined politically and defined according to the sector on a case-by-case basis, whereby the following principles apply:

Content: what is included in basic service provision must be specified by legislation.

Need: the needs of the population (like those of companies) will change. This change must be taken into account.

Accessibility and principle of area coverage: the services must be provided according to the same principles and to all. They must be provided in all regions of the country and must be easily accessible for all segments of the population. The application of this rule contributes to social and regional cohesion.

Quality: the services must be of good quality. The quality requirements are stipulated in laws and regulations and are monitored and enforced by public authorities.

Price: the prices for the services must be affordable for all. The application of this rule contributes to economic and social cohesion.

Continuity: the services must be provided without interruption. Their continuous provision is usually also in the business interest of the service provider. It is of particular importance in times of crisis (including to safeguard the country's security interests) (translation, original quote in German)."

---

263 NFA Communication, BBl 2002, 2459.

264 Report by the Federal Council "Grundversorgung in der Infrastruktur", BBl 2004, 4569, 4578.

Which services are considered basic services is a *political decision* and depends on the opinions and changes of the time. Art. 43a par. 4 BV consequently does not contain any definition of the term "basic service provision" in terms of content but is worded openly in this regard. Even so, these are expected to be "sensitive areas"[265]. Although the clause, just like the social objectives (art. 41 BV), is not actionable, i.e., it does not give rise to any enforceable claims[266], it is however aimed at the legislature and public authorities, namely those of both the Federation and of the cantons[267]. In addition, conclusions can be indirectly drawn in the sense that services are involved that are so important—even sensitive—that they are accessible (affordable) for everyone and should be provided spatially inclusively as well as continuously.

Statutory building insurance against natural hazards represents a "basic service provision" or a "public service" from today's point of view, all the more so in the light of climate change. There is an eminent public interest in this service being provided in such as way as to be accessible to all, spatially inclusive and continuous. The statutory insurance system of the KGV is therefore supported by art. 43a par. 4 BV as well.

In comparative law, it is conspicuous that the term "basic service provision" more or less corresponds to the terms coined by EU law "provision of general interest services", "universal service" as well as "services of general (economic) interest" (see p. 118 et seq. below).

5. The SUVA as a comparable example of public service

The public service has been strengthened in recent years within the framework of legislation as well. An example that is particularly interesting for statutory insurance against natural hazards is cited below.

The Swiss Accident Insurance Law (Unfallversicherungsgesetz, UVG) is currently being revised[268]. Within the framework of the current UVG, the SUVA[269], an independent institution under public law, has a legal partial monopoly for occupational and non-occupational accidents. Certain categories of employees have to take out accident insurance from the SUVA. This is therefore a case of statutory insurance provided by the SUVA. Mainly employees working in accident-prone branches of

265 NFA communication, BBl 2002, 2459; SCHWEIZER/MÜLLER, St. Galler Kommentar BV, BV 43a N 22.
266 NFA communication, BBl 2002, 2459.
267 RHINOW/SCHEFER, p. 127.
268 Federal Law on Accident Insurance (UVG) dated 20 March 1981 (SL 832.29). Cf. UVG communication, BBl 2008, 5395 et seq.
269 Schweizerische Unfallversicherungsanstalt (Swiss Accident Insurance Institution); cf. art. 58 et seq. UVG.

industry are included in this, e.g. from the construction and the forestry industry or the metalworking industry[270].

On the occasion of the revision, a controversy was argued regarding the preservation of the partial monopoly. Based on miscellaneous economic studies, the SFC concluded that the partial monopoly is efficient and advantageous and rejected privatisation[271]. A regulatory debate was conducted in the press[272]. The Federal Office of Justice commented in detail regarding the issue of constitutionality of the partial monopoly in a legal opinion for the attention of the responsible, pre-advisory commission of the National Assembly on 29 September 2008 and concluded in a convincing statement of grounds that the partial monopoly is compatible with the BV, in particular with economic freedom[273]:

"Prior to revision of the current law, the following reasons have been asserted for the partial monopoly of the SUVA:

- *The SUVA provides insurance according to the principle of reciprocity (art. 61 par. 2 UVG), that is, it does not aim at profit. Owing to its monopoly, it requires only a small amount of equity capital and in addition does not have to pay interest on this. Also, no acquisition costs are incurred in the area covered by its monopoly. If the partial monopoly were to be abolished, the SUVA would have to be privatised. Interest would have to be paid on the equity capital (at its current level). In the cost-benefit analysis by the University of St. Gallen (Prof. Franz Jäger, October 2004), these costs are estimated at CHF 60 million (no. 5.2.4); these would have to be reabsorbed by the premium payers. Acquisition costs, which would be added to the premiums, would be incurred by the SUVA on the open market as well. In the cost-benefit analysis (no. 5.3.4), these costs are estimated at CHF 24–32 million. With an eye to these specifics and based on a detailed output/input relation, the cost-benefit analysis (no. 6.2.8) therefore arrived at the conclusion that, compared to private insurers, the SUVA paid a greater share of its revenues back to the policyholders again through its insurance benefits.*
- *Considering that the accident risks are greater in the SUVA industry sectors than with other industry sectors insured by other UVG insurers, it is apparent that the financial advantages of the SUVA insurance (no pursuit of profit, no interest paid on equity capital, no acquisition costs) make it possible to prevent the sector- or risk-related differences in premiums between undertakings that are SUVA-*

270 Art. 66 UVG.

271 UVG communication, BBl 2008, 5401 et seq.

272 Cf. NZZ dated 26 May 2009, p. 21: "Triftige Gründe für den Status quo" ("Compelling reasons for the status quo") by Marie-Thérèse Weber-Gobet versus "Valable Argumente für eine Privatisierung" ("Valid arguments for privatisation") by Franz Jäger.

273 Expert report BJ on UVG revision, particularly p.4 et seq.

insured and those that are insured with other UVG insurers from getting too high. Hence, the partial monopoly has rather a compensating effect with regard to the insurance premiums, and this may by all means be desirable in social insurance (cf. Duc, loc. cit.).

- *Similar considerations apply with regard to the individual industry sectors insured with SUVA. Although even SUVA is obliged to determine its premiums in a risk-adjusted manner (art. 92 UVG), owing to the partial monopoly, the industry sector-specific risks (e.g. with respect to accidents in all carpenter's workshops) are better balanced than when undertakings operating in the same sector of industry take out insurance from several private insurance companies, because this results in a reduction of the risk community. Even the sector-related review thus leads to lower premiums as a result, and this is in the interest of the insured undertakings. What has been said above also applies to the non-industrial accident insurances. In accordance with art. 92 par. 6 UVG, tariff classes can be configured for these, with sex-related differentiation being inadmissible. This should not be changed even in the revised version of art. 92 requested by the Federal Council in the communication on the UVG revision.*
- *As is shown in the cost-benefit analysis (no. 6.2.8, bottom of p. 78 et seq.), there are also no indications that potential efficiency disadvantages of the SUVA could cancel out the financial advantages of the partial monopoly. To wit, even today, the SUVA is participating in a "potential competition": its benefits and premiums could be compared with those of private insurers; employers are represented in the administrative board of the SUVA and could resist premium increases; the sword of Damocles of liberalisation and privatisation is a disciplinary measure to maintain the pressure on the SUVA to increase efficiency.*
- *Although the abolition of the partial monopoly of the SUVA would make additional business opportunities available to the private insurance sector, undertakings that have an increased risk of occupational accidents and occupational illnesses would be burdened more severely as a result. This would especially affect the premiums for non-industrial accident insurance (cf. no. 3.8. of the cost-benefit analysis). If need be, a premium price-reduction system would have to be introduced that would burden the government with about CHF 200–300 million (no. 6.8 of the analysis, p. 104). In addition, provisions and reserves of the SUVA would have to be raised massively prior to privatisation in order to capitalise the future deliverables of the SUVA, namely the cost of living bonuses on existing retirement benefits (they are financed nowadays primarily based on a pay-as-you-go system). This could cost the Federation up to CHF 4 billion (cf. no. 5.1 of the analysis).*

The present remarks show that a substantial number of noteworthy reasons can be propounded in support of the partial monopoly of the SUVA. On the one hand, the partial monopoly has clear advantages for SUVA-insured industries. On the other hand, it is a cost-effective solution for the Federal Treasury as well (and thus for the taxpayer).

On the whole, the system contributes to balancing the financial consequences of the various occupational and non-occupational accident risks of the various industries with higher risks. From this point of view, the legislature, in our opinion, does not offend against equal rights if it treats employers and employees from various industries differently based on the partial monopoly of the SUVA.

1.2.7 Based on the above observations, the public interests justifying the deviation from the economic freedom of private insurance companies can now also be identified. On the one hand, it is the socio-political interest in maintaining a certain degree of solidarity in accident insurance. This interest is of great importance in any social insurance (cf. Duc, loc. cit., margin no. 22, with additional references). On the other hand, fiscal interests are involved as well to a certain extent, for the suspension of the partial monopoly would place a considerable burden on the Federal Treasury. These two interests must be balanced against the interests of the private insurance companies in additional business opportunities. In our view, it is thoroughly acceptable in this weighing of interests to give priority to the socio-political interest in solidary accident insurance and in sparing the federal treasury as opposed to the commercial interests of the private insurance companies. The maintenance of the partial monopoly can also be justified in terms of proportionality for there is—as far as we can see—no softer means of achieving the target objectives. It should be emphasised that taking into account fiscal interests in the present case has nothing to do with a (in principle inadmissible) fiscal exploitation of a monopoly (cf. Ehrenzeller, St. Gallen Kommentar; 2nd ed. 2008, on art. 27, margin no. 73 et seq.; Rhinow/Schmid/ *Biaggini, Öffentliches Wirtschaftsrecht, 1998, § 18 margin no. 69 et seq.). Such a case would result if the SUVA had an absolute monopoly and if it misused it to make a profit at the expense of the policyholders by charging exorbitant premiums. There can be no question of this (translation, original quote in German)."*

Although the revision is still underway, it can be expected that the partial monopoly will be maintained. The parallels with statutory insurance against natural hazards are obvious. Even the SUVA's partial monopoly is a monopoly that is justified for socio-political reasons. All employees should have access to accident insurance at affordable premiums. This requires a solidary insurance that balances good and bad risks. This in turn can be done only on the basis of a legal monopoly that ensures a sufficiently large risk collective. The SUVA as well as the KGV statutory insurance represents a "basic service provision" or a public service.

6. Conclusion

The public-law statutory building insurance against natural hazards, as provided by the KGV, is accessible to all building owners at affordable premiums and guarantees comprehensive and complete insurance protection. It is inevitably based on the principle of solidarity which balances good and bad risks. For such insurance protection to be granted at all is in the public interest. This will become even more important when faced with climate change. The legal monopoly of the statutory insurance is justified for socio-political and police reasons, as the target objective, solidary insurance, can be achieved only on the basis of the monopoly. The integration of damage prevention and damage abatement as well as the synergy effects reinforce solidarity, and for these reasons the monopoly is justified for police reasons as well. The statutory insurance system is therefore consistent with economic freedom and the KG. Furthermore, it can be qualified as a basic service provision within the meaning of art. 43a par. 4 BV which must be accessible to all persons at comparable conditions, in other words, affordable premiums and appropriate terms, based on the constitution. The KGV statutory insurance system against natural hazards is thus legally permissible. The SUVA example shows that there are also other statutory insurance systems whose monopoly is legally permissible and that are generall acknowledged to represent a public service.

7. Excursus: system competition as the third method?

A. System competition between the KGV and private insurances

System competition generally refers to the competition among legal systems, frequently against the background of location attractiveness for investors or the debate on the best regulation (competition among fiscal systems, competition in company law). This competition can also exist in the rivalry between different competitive orders, namely one that is characterised by a legal monopoly and one that is characterised by free market economy.

With regard to insurance against natural hazards, a statutory market and price system in line with the KG exists both in the cantons that have the KGV as well as in the GUSTAVO cantons. There are, however, significant differences in the configuration of these systems, as there is only *one* provider subject to public law on the side of the KGV, which has a legal monopoly, while several providers are operating under private law in the GUSTAVO cantons (cf. p. 46 et seq. and 64 et seq.). One can therefore speak of system competition even in the sector that provides insurance against natural hazards.

B. Positive effect of system competition

The positive effect of system competition has been emphasised by several parties, including by the Competition Commission in the context of building insurance, as well as, for example, regarding the revision of the UVG (the SUVA has only a partial monopoly which is why its accident insurance benefits can be compared with those of private insurances). The positive effect is manifested as follows[274]:

- Due to their monopolies, the KGV are constantly being "monitored" to a certain degree. Their services are compared with those provided by private insurances companies. Although the pressure to give up the monopolies or to "deregulate" has decreased in recent years, it is never going to disappear completely. In order to be able to withstand this pressure and to be able to persist in comparison with private insurances, the KGV will always have to perform impeccably.
- On the other hand, the services provided by private insurance companies are compared with those of the KGV. Thus, the Federal Court and the Competition Commission have found that, *overall,* the premiums of the KGV are significantly (about 40%) lower than those of private insurances[275]. It is legally possible to introduce a new, cantonal legal building insurance monopoly in a GUSTAVO canton[276]. Private insurances are therefore under the sword of Damocles of the introduction of a new KGV monopoly. They must therefore provide good services as well or they come under pressure to improve their services.
- Both the KGV and private insurance companies are continuously disciplined by system competition. They must show that they are better in terms of quality and price of service (claims settlement) than the service providers in the other system.
- Hence, a certain degree of competition does indeed take place in terms of quality and price through competition between the two different systems.

Based on system competition, the Competition Commission concluded, last but not least, that neither a recommendation to "deregulate" nor to introduce new KGV monopolies is appropriate with regard to building insurance[277]. One can definitely concur with this. System competition comprises an advantage that should not be underestimated, which would be lost if the service were provided by one type of system only. In addition, by consciously upholding and committing to system competition,

---

274 Recht und Politik des Wettbewerbs (RPW) 2003, p. 752; regarding SUVA see p. 103 et seq. above; Brinkmann, p. 37 et seq.

275 FCR 124 I 25, 29 et seq. as well as Recht und Politik des Wettbewerbs (RPW) 2003, p. 741 et seq.; cf. p. 92 above.

276 To do so, the requirements of public interest, adequate legal basis and proportionality must be met. The first and the latter, as explained on p. 90 et seq. and 93 et seq., exist, and an adequate legal basis can be created by issuing a cantonal law to that effect.

277 Recht und Politik des Wettbewerbs (RPW) 2003, p. 752.

the "controversy over beliefs", which is in itself unproductive, can be avoided and the required comparison of systems can be limited exclusively to the factual level.

Against this background, system competition and thus the coexistence of two systems (building insurance with and without a legal monopoly) indeed appears to be a third method worth considering.

II. EU compatibility of the KGV statutory insurance system

1. EU compatibility as a permanent issue for the Swiss legislature

Switzerland is not a member of the EU and for the time being it is also not be expected that it will become a member any time soon. The question why EU compatibility of the statutory insurance system of the KGV can even be a relevant subject is therefore legitimate. The EU is Switzerland's most important trade partner by far; after all, the country is completely surrounded by the EU from a purely geographical point of view[278]. This, and the extremely strong interlacing of the economic areas, has resulted in each Swiss legislative proposal, regardless of what level, being systematically examined with respect to its EU compatibility. By default, each draft bill nowadays accounts for its EU compatibility. Frequently, legislative proposals are even triggered by a change in EU laws, such as the complete revision of the Swiss Insurance Supervision Law (VAG). Thus, the communication to that effect by the SFC states: *"Along with the reorientation of supervision, a second objective of the revision—the improvement of EU compatibility of Swiss law—should be achieved (translation, original quote in German)."*[279]. In the meantime, one can refer to a *systematic reproduction* of EU law[280].

This circumstance can result in existing Swiss laws (at the federal or the cantonal level) being challenged due to their alleged lack of EU compatibility and a revision being proposed. However, this does not mean that Swiss law is inevitably adjusted to EU law. An adjustment is often not carried out or only in part, because EU law is just

278 In the direct insurance sector (without life insurance), Switzerland and the EU have granted each other mutual access to the market by entering into a bilateral agreement; foreign insurance companies can set up a branch office in each case (Agreement between the Swiss Confederation and the European Economic Community concerning direct insurance other than the life insurance dated 10 October 1989, SL 0.961.1). In the area of responsibility of the KGV, access to the market is, however, explicitly prohibited (art. 3 in conjunction with Annex no. 2(D) Agreement). This is designed to prevent "cherry-picking" by the foreign insurance company which would endanger the stability of the KGV system which is oriented to the common good.

279 VAG communication 2006, BBl 2003, 3789, 3795.

280 For fundamental information SPINNER/MARITZ, in: FS Zäch, p. 127 et seq.

not considered to be the most practicable or best solution or an adjustment simply turns out to be unnecessary. Furthermore, even though Switzerland is not a member, the question is occasionally raised as to whether the Swiss regulation could be maintained, should Switzerland join the EU, or whether an adjustment would be required at that point in time. Although it is unnecessary per se, Swiss law is sometimes, practically "for stockpiling", adapted to EU law on the basis of such considerations. This is partly done under the heading of "modernisation" of Swiss law, and in the course of such adaptions, the issue of how a certain sector is regulated in the EU is periodically re-examined. Knowledge of the EU law pertaining to the sector concerned is, in any event, indispensable due to this starting situation.

In addition, even independent of the Swiss situation, the question arises as to whether a statutory insurance system is compatible with EU law. Due to climate change, the establishment of such systems will inevitably become an issue even in the EU region. It has been shown that statutory insurance is the appropriate and necessary response to climate change. The question of EU compatibility is therefore going to arise for the EU Member States as well.

The extent to which *this type* of statutory insurance system against natural hazards is compatible with EU law will be examined below *using the example of the KGV.* The point here is to clarify whether such a statutory insurance system can endure within the framework of EU law or whether fundamental incompatibilities exist in this regard.

2. Conflicting priorities: internal market—competition—common good

On the one hand, the EU considers itself a market without (internal) borders, which is innovative and competitive. The EU internal market is safeguarded by the fundamental freedoms—the free movement of goods, the freedom of establishment, the free movement of services and the free movement of capital. Competition, which represents another principle of the internal market, is intended to protect freedom to conduct a business from intervention and facilitate the undistorted competition in trading among the Member States[281].

On the other hand, the EU has increasingly developed from a community with focus on a uniform economic area into a community of values over the last 10–15 years. EU law goes far beyond the scope of economic regulations nowadays.

The following sectors of conflicting priorities are relevant in evaluating the EU compatibility of the statutory insurance system:

281 OPPERMANN/CLASSEN/NETTESHEIM, p. 363 et seq., 414 et seq.

- Free movement of services versus the exercise of official authority
- Free movement of services versus common good
- Competition versus common good

The EU—not least by means of ECJ case law[282]—has created a system of rules to resolve potential conflicts and contradictions. These rules and other provisions may serve as a basis for the statutory insurance system.

3. Basis for EU compatibility of the statutory insurance system—an overview

The statutory insurance system may be based on the following fundamental principles:

- Statutory insurance as the exercise of official authority (see p. 108 et seq. below);
- Statutory insurance as the expression of overriding public interest / compelling general good (see p. 112 et seq. below);
- Statutory insurance as a facility for the provision of general interest services, or as a service of general (economic) interest (see p. 118 et seq. below);
- Statutory insurance as a permissible compensation scheme in law governing State aid (see p. 134 et seq. below);
- Statutory insurance as a measure for implementing civil protection (see p. 138 et seq. below).

The different fundamental principles will be discussed in detail below, based on the regulations in the **Treaty of Lisbon**, namely the *"Treaty on the European Union* ("TEU") and the *Treaty on the Functioning of the European Union* ("TFEU") dated 13 December 2007[283]. Both treaties entered into force on 1 December 2009. The latter treaty had thus far been called *Treaty establishing the European Community*, EC treaty for short ("TEC"). The Treaty of Lisbon effected a change in the article numbering compared to the Treaty of Nice[284] which had been valid until then. Since the previous case law and theory still refers to the numbering according to the version of

---

[282] Court of Justice of the European Union, European Court of Justice for short, based in Luxembourg.

[283] Treaty of Lisbon amending the Treaty on European Union and the Treaty establishing the European Community, signed in Lisbon on 13 December 2007, OJEU C306 dated 17 December 2007, p. 1 et seq. A consolidated version of the Treaty on the European Union and the Treaty on the Functioning of the European Union can be found in OJEU C115 dated 9 May 2008, p. 1 et seq.

[284] Treaty of Nice amending the Treaty on European Union, the Treaties establishing the European Communities and certain related acts dated 26 February 2001, OJEU 2001/C 80 p. 1 et seq. A consolidated version can be found in OJEU 2002/C 325 p. 5 et seq.

the Treaty of Nice, the article number according to the Treaty of Nice will be given in square brackets in each case.

4. Statutory insurance as the exercise of official authority

A. Starting situation: neither a substantive privatisation obligation nor a monopoly ban based on EU law

According to art. 345 TFEU [295], *"the Treaties shall in no way prejudice the rules in the Member States governing the system of property ownership."*

The different significance and the different extent of the public sector in the Member States, particularly in the early stage of the community, are hidden behind this clause[285]. It means that it is solely up to the Member States to decide on whether a certain activity shall be performed by an undertaking subject to public law or an undertaking subject to private law. In particular, the Member States are free to decide on whether to nationalise or to privatise an undertaking[286]. The EU, by contrast, has no power of privatisation and the Community Law is neutral regarding the allocation to public or private ownership. Determining the economic constitution is the responsibility of the Member States and consequently they determine whether a certain task is performed by the state or by the private sector[287]. Therefore, no substantive privatisation obligation at all ensues from EU law. This, however, does not mean that a public undertaking is automatically exempted from the EU fundamental freedoms and competition law[288]. For that to take place, additional requirements must be met. All the same, it has been made clear though that an insurance system *regulated by public law* is compatible with EU law.

In addition, it should be emphasised that there is no regulation in EU law that prohibits a service monopoly regulated by public law. Instead, EU law assumes the existence of such monopolies, and furthermore there is also no requirement that such monopolies be abolished or revised (the latter is in contrast to trade monopolies[289]). *EU law is thus based on the principle that service monopolies regulated by public law are permitted*[290]. If, exceptionally, such a monopoly is not lawful, this will only emerge from an examination on a case-by-case basis.

---

[285] Hatje, in: Schwarze, EGV 295 N 1.

[286] Hatje, in: Schwarze, EGV 295 N 2 et seq.

[287] Kingreen/Wegener, in: Calliess/Ruffert, EGV 295 N 10 et seq.

[288] Kingreen/Wegener, in: Calliess/Ruffert, EGF 295 N 12.

[289] Cf. art. 37 TFEU [31].

[290] Kluth, in: Calliess/Ruffert, EGV 49/50 N 88 et seq.; Emmerich, in: Dauses, vol. 2, book II N 72; Berg, in: Schwarze, EGV 31 N 6.

B. No application of the free movement of services in the exercise of official authority

Pursuant to art. 56 par. 1 TFEU [49 par. 1], *"within the framework of the provisions set out below, restrictions on freedom to provide services within the union shall be prohibited in respect of nationals of Member States who are established in a Member State other than that of the person for whom the services are intended."* The term "nationals" also refers to companies[291].

This provision incorporates the free movement of services ("FMS"), one of the fundamental freedoms of the EU. It contains, among other things, an insurance company's right to provide cross-border services, that is, to conclude cross-border insurance contracts, without a branch office within the state where the policyholder resides (free movement of correspondence services)[292].

Pursuant to art. 62 in conjunction with art. 51 TFEU [55/45], the FMS does not apply from the outset, however, if the activity concerned is *"connected, even occasionally, with the exercise of official authority"* in a member state.

Acting in a manner subject to public law does not yet represent the exercise of official authority. What is required is direct and specific participation in the exercise of official authority. This must comprise the possibility to exercise coercive power. Acting under sovereign jurisdiction is required. In addition, the exercise of official authority must be necessary for the purpose. If justice can also be done to public interests by means of less drastic measures such as authorisation requirements or a supervisory authority, the FMS shall have priority[293]. According to case law, bailiffs and in some cases notaries, for example, exercise official authority, although not expert witnesses for traffic accidents who appear in court, or auditors for insurance companies[294]. The prerequisites show that it should not be possible without further ado to remove entire sectors from FMS by appealing to official authority. On the other hand, national idiosyncrasies should be taken into account to a certain degree here as well, and the Member States are therefore accorded a certain margin of appreciation as to which sectors they would like to assign to official authority[295].

If official authority is exercised, this will simultaneously also lead to non-application of the EU competition law for it lacks the undertaking element, that is to say, the element of economic activity (see also p. 121 et seq. below).

---

291 Art. 62 TFEU [55] in conjunction with art. 54 TFEU [48].

292 Oppermann/Classen/Nettesheim, p. 467.

293 Bröhmer, in: Calliess/Ruffert, EGV 45 N 2 et seq., 6, 8 et seq.; Holoubek, in: Schwarze, EGV 49/50 N 44.

294 Bröhmer, in: Calliess/Ruffert, EGV 45 N 3, 10; Holoubek, in: Schwarze, EGV 49/50 N 44.

295 Schlag, in: Schwarze, EGV 45 N 5.

The above principles also apply to the freedom of establishment ("FOE"), another fundamental freedom of the EU that grants insurance companies the right to set up a branch office in the broadest sense of the word in another member state and operating there permanently[296].

The KGV exercise official authority in the sense mentioned above. They have coercive powers at their disposal by being able to deny approval for a construction project, by ordering that property protection measures be taken, and by enforcing insurance coverage through the KGV. The coercive powers are partly a consequence of the integration of the prevention of damage caused by natural hazards (see p. 56 et seq., 58 et seq. above). In addition, the question arises whether the execution of such coercive powers is necessary in order to fulfil the public purpose, i.e., comprehensive insurance against natural hazards that is accessible to all. The remarks on p. 56 et seq., 90 et seq. and 93 et seq. have shown that the integration of prevention is necessary in order to strengthen the solidary insurance system and to preserve it in the long run. Likewise, statutory insurance in itself, and the constraint associated with it, is indispensable for safeguarding the required risk collective and thus solidary, comprehensive insurance. In view of the margin of appreciation that the Member States have, the exercise of official authority, which has priority over FMS, can be assumed.

Hence, the FMS, FOE and EU competition law already do not apply based on the fact that the KGV exercise official authority. This constitutes the first foundation for EU compatibility of a statutory insurance system such as that of the KGV.

C. The Consorcio de Compensación de Seguros solution—qualification as tax

The Consorcio also represents a statutory insurance. It is domiciled in Spain and is thus located in a EU member state. The foundation on which the Consorcio is based as regards EU compatibility will be demonstrated briefly below.

As demonstrated on p. 42 et seq., the mandatory surcharge raised by the Consorcio safeguards its monopoly since, on payment thereof, insurance protection through the Consorcio is automatic and there is no need for additional (private) insurance coverage. The surcharge regulation is nothing other than the establishment of a legal monopoly.

It was subsequently demonstrated that this surcharge qualifies as tax under Spanish law (see p. 43 above). This qualification *has been transferred to EU law.* Thus, art. 46

296 Art. 49 and 51 TFEU [43/45]. Oppermann/Classen/Nettesheim, p. 520 et seq., 524.

par. 2 of Directive 92/49[297], by which free movement of services for bulk business in the direct insurance sector was introduced, states:

"Without prejudice to any subsequent harmonization, every insurance contract shall be subject exclusively to the indirect taxes and parafiscal charges on insurance premiums in the Member State in which the risk is situated as defined in Article 2 (d) of Directive 88/357/EEC, and also, in the case of Spain, to the surcharges legally established in favour of the Spanish 'Consorcio de Compensación de Seguros' for the performance of its functions relating to the compensation of losses arising from extraordinary events occurring in that Member State.

In derogation from the first indent of Article 2 (d) of Directive 88/357/EEC, and for the purposes of this paragraph, moveable property contained in a building situated within the territory of a Member State, except for goods in commercial transit, shall be a risk situated in that Member State, even if the building and its contents are not covered by the same insurance policy.

[297] Council Directive dated 18 June 1992 on the coordination of laws, regulations and administrative provisions relating to direct insurance other than life assurance and amending Directives 73/239/EEC and 88/357/EEC (third non-life insurance Directive), 92/49/EEC, OJEU L228 dated 11 August 1992, p. 1 et seq., hereinafter referred to as Directive 92/49. Directive 92/49 will be repealed and displaced as of 1 November 2012 by the new Directive 2009/138/EC (Directive 2009/138/EC by the European Parliament and the Council dated 25 November 2009 regarding the taking up and pursuit of business Insurance and Reinsurance (Solvency II), OJEU L335 dated 17 December 2009, p. 1 et seq., hereinafter referred to as Directive 2009/138; cf. art. 310 Directive 2009/138). The qualification of the surcharge as tax and thus the preservation of the monopoly of the Consorcio is, however, also safeguarded by the new Directive 2009/138. Thus, even recital 87 of the same already contains a provision in favour of the Member States' fiscal systems and the new art. 157, "Taxes on premiums", states:

"1. Without prejudice to any subsequent harmonisation, every insurance contract shall be subject exclusively to the indirect taxes and parafiscal charges on insurance premiums in the Member State in which the risk is situated or the Member State of the commitment.

For the purposes of the first subparagraph, movable property contained in a building situated within the territory of a Member State, except for goods in commercial transit, shall be considered as a risk situated in that Member State, even where the building and its contents are not covered by the same insurance policy.

In the case of Spain, an insurance contract shall also be subject to the surcharges legally established in favour of the Spanish 'Consorcio de Compensación de Seguros' for the performance of its functions relating to the compensation of losses arising from extraordinary events occurring in that Member State.

2. The law applicable to the contract under Article 178 of this Directive and under Regulation (EC) No 593/2008 shall not affect the fiscal arrangements applicable.

3. Each Member State shall apply its own national provisions to those insurance undertakings which cover risks or commitments situated within its territory for measures to ensure the collection of indirect taxes and parafiscal charges due under paragraph 1."

Although Directive 2009/138 already entered into force on 6 January 2010 (art. 311 par. 1), art. 157, however, like the other existing relevant provisions, will not enter into force until 1 November 2012 (art. 311 par. 2). Other provisions of the above-mentioned Directive must be implemented by national legislation as of 31 October 2012 (art. 309 par. 1).

The law applicable to the contract under Article 7 of Directive 88/357/EEC shall not affect the fiscal arrangements applicable.

Pending future harmonization, each Member State shall apply to those undertakings which cover risks situated within its territory its own national provisions to ensure the collection of indirect taxes and parafiscal charges due under the first subparagraph."

This provision is designed to stipulate that, on a fire insurance contract, for example, between an insurance company domiciled in Germany and the owner of a building, the surcharge for his house in Spain will have to be paid in any event, whereby the building is automatically insured against natural hazards with the Consorcio. Even within the framework of the free movement of correspondence services, the Consorcio's monopoly thus cannot be undermined. In addition, subparagraph 4 of the provision indicates that such fiscal arrangements are the responsibility of the Member States and do not come under the jurisdiction of the EU[298].

The fiscal regime shows that the Consorcio exercises official authority. Surcharge and insurance coverage can be enforced compulsorily by the Consorcio. The surcharge represents a claim deed regulated by public law and can be enforced directly (see p. 43 above). The fiscal regime is required in order to safeguard comprehensive solidary insurance coverage that is accessible to all. Hence, FMS, FOE and even EU competition law do not apply.

Through a relatively simple regime, the Consorcio has prevented the removal of the monopoly regulation to this day, in spite of the provision of art. 3 Directive 92/49 in conjunction with art. 4 Directive 73/239[299] which provides for the repeal of

---

298 Only EU regulations that are, above all, designed to prevent discrimination against foreign persons or deliverables (cf. art. 110 et seq. TFEU [90 et seq.] are reserved. This legal situation will not change even as a result of the new Directive 2009/138, cf. footnotes 297 and 299.

299 The first Council Directive dated 24 July 1973 on the coordination of laws, regulations and administrative provisions relating to the taking up and pursuit of the business of direct insurance other than life assurance) (first non-life insurance Directive), 73/239/EEC, OJEU L228 dated 16 August 1973, p. 3 et seq., hereinafter referred to as Directive 73/239. Directive 73/239 will also be repealed as of 1 November 2012 by the new Directive 2009/138 (cf. footnote 297). Even in art. 188 of the latter, the repeal of certain monopolies of institutions explicitly listed in the Directive is upheld as before. Art. 188 in conjunction with art. 8 Directive 2009/138, however, is merely the unaltered continuation of art. 3 Directive 92/49 in conjunction with art. 4 Directive 73/230. The Consorcio had already been mentioned in art. 4 Directive 73/239 (cf. consolidated version, Celex no. 01973L0239-20070101) and now continues to be mentioned in the "succession article" 8. However, the proviso of a national fiscal regime in accordance with art. 157 Directive 2009/138 is still not affected by this and, apart from that, the special configuration of the Consorcio is protected by art. 8 which exempts the Consorcio from the scope of the new Directive as before:
"This Directive shall not apply to the following institutions which pursue non-life insurance activities unless their statutes or the applicable law are amended as regards capacity:
[...]
4. in Spain, the Coonsorcio de Compensación de Seguros."

insurance monopolies by 1 July 1994. The German public-law building insurance providers have abolished their monopoly as of that date and have not introduced any regulation corresponding to the Consorcio. This is explained by the fact that the German building insurers were not prepared to ultimately take a stand for the continued existence of the monopolies and did not have the required political support[300].

The qualification as tax also means that the "service" element is absent. A service for the purpose of the FMS must, by definition, be in return for payment. If the service is financed by a public charge, in other words, a tax, there is no payment basis and thus no service[301]. Hence, again, the FMS does not apply.

The example of the Consorcio as an existing statutory insurance within the EU provides proof that statutory insurance is compatible with EU law. The chosen solution stands to reason and is therefore all the simpler.

5. Statutory insurance as an expression of overriding public interest

A. The broad term of overriding public interest

As grounds of justification[302], case law has developed the term "overriding public interest" for a *restriction* of FMS. This is comparable with public interest in Swiss law. What is comprised by this is naturally not conclusively defined but instead depends on the social development and is subject to the change of opinions over time[303]. Up to now, the ECJ has subsumed a wide variety of interests under this term, from consumer protection and health protection through to the protection of non-commercial broadcasting[304]. The general good of *"protection of the social order"* (in accordance

---

This ensures that the mandatory surcharge and the prescribed coverage clause which constitute the legal monopoly continue to be preserved in spite of art. 188 and its reference to art. 8, even under the new Directive. Hence, the legal monopoly of the Consorcio will continue to exist, unchanged, even after 1 November 2012 (art. 8 and 188 Directive 2009/138 will enter into force on 1 November 2012; regarding the personal insurance sector covered by the Consorcio, cf. art. 10(3) Directive 2009/138). Furthermore, recital 4 of Directive 2009/138 makes it clear that special insurance organisations such as the Consorcio and special insurance coverage will continue to remain unaffected even in the future: *"It is appropriate that certain undertakings which provide insurance services are not covered by the system established by this Directive due to their size, their legal status, their nature—as being closely linked to public insurance systems—or the specific services they offer. It is further desirable to exclude certain institutions in several Member States, the business of which covers only a very limited sector and is restricted by law to a specific territory or to specified persons."*

300 VON UNGERN-STERNBERG, p. 154 et seq.

301 KLUTH, in: CALLIESS/RUFFERT, EGV 49/50 N 12; HOLOUBEK, in: SCHWARZE, EGV 49/50 N 20.

302 This is in contrast to the exercise of official authority which does not represent grounds of justification but instead leads to an activity being outside of the scope of application of the FMS. HOLOUBEK, in: SCHWARZE, EGV 49/50 N 44.

303 Cf. HOLOUBEK, in: SCHWARZE, EGV 49/50 N 110.

304 HOLOUBEK, in: SCHWARZE, EGV 49/50 N 111 et seq., with numerous references to case law.

with socio-cultural particularities, that is to say, according to the conventions of the state concerned) and *"prevention of the disturbance of the social order in general"*, which are recognised by the ECJ, are particularly broad and therefore in need of concretisation[305].

The specific restriction can reside in a legal monopoly. Thus, the ECJ has recently protected the legal monopoly of a non-profit legal entity regulated by public law in Portugal ("Santa Casa") for lottery, sports pools and all types of sports bets. The monopoly was declared legitimate with reference to overriding public interest and Bwin, an online vendor of gambling games, was prohibited from carrying out any lotto, pools and other gambling game activity on the internet for the national territory of Portugal. The ECJ judged that Bwin cannot invoke FMS which may be restricted on the grounds of overriding public interest[306].

B. Considerable threat to the financial equilibrium of a social security system as overriding public interest

In the year 2000, the EU Commission has published an *"Interpretative communication"* in the insurance industry entitled *"Freedom to provide services and the general good in the insurance sector"*[307]. This is aimed especially at the responsible public authorities in the Member States and is designed to promote a uniform understanding of terms. It is, however, not binding, least of all for the ECJ. The communication itself states in this regard: *"It goes without saying that the Commission's interpretations do not prejudge the interpretation that the Court of Justice of the European Communities, which is responsible in the final instance for interpreting the Treaty and secondary legislation, might place on the matters at issue."*[308].

In the above-mentioned communication, various points, such as the language of the insurance contract or uniform bonus-penalty systems are identified as expressions of general good that allow restrictions of the FMS. The communication has become obsolete in the meantime as the result of the case law of the ECJ. In recent years, the ECJ has recognised another overriding public interest, that of *"the risk of seriously undermining the financial balance of the social security system"*[309] (cf. also p. 121 et seq. below in this regard on competition law).

---

305 Holoubek, in: Schwarze, EGV 49/50 N 113; ECJ case C-42/07 in the matter of Bwin versus Santa Casa, judgment of 8 September 2009, marginal no. 56. ECJ case C-338/04, C-359-04 and C-360/04 in the matter of Placanica, judgment of 6 March 2007, marginal no. 46.

306 ECJ case C-42/07 in the matter of Bwin versus Santa Casa, judgment of 8 September 2009.

307 Hereinafter referred to as Insurance Sector Interpretative Communication.

308 Insurance Sector Interpretative Communication, introduction, paragraph 11.

309 ECJ case C-385/99 in the matter of Müller-Fauré, judgment of 13 May 2003, marginal no. 73; Holoubek, in: Schwarze, EGV 49/50 N 112.

In a legal dispute that was fought recently, the legitimacy of the legal monopoly of a German workmen's compensation insurance, that is to say, the legitimacy of a statutory insurance, was to be evaluated. The Kattner Stahlbau GmbH (Kattner), a company in the metalworking company, refused to insure its personnel with the Maschinenbau- und Metall-Berufsgenossenschaft (MMB), a workmen's compensation insurance comparable to the Swiss SUVA, which has a legal monopoly. Kattner wanted to insure its personnel with a different, private insurer and claimed that the monopoly violated the EU competition law and the FMS. The ECJ ruled as follows[310]:

"Consequently, the answer to the first question referred for a preliminary ruling must be that Articles 81 EC and 82 EC[311] are to be interpreted to the effect that a body such as the employers' liability insurance association at issue in the main proceedings, to which undertakings in a particular branch of industry and a particular territory must be affiliated in respect of insurance against accidents at work and occupational diseases, ***is not an undertaking within the meaning of those provisions, but fulfils an exclusively social function, where such a body operates within the framework of a scheme which applies the principle of solidarity and is subject to State supervision****, which it is for the referring court to verify." [...]*

"In this respect, it must be noted that, according to the Court's case-law, the ***risk of seriously undermining the financial equilibrium of the social security system may constitute an overriding reason in the public interest capable of justifying an obstacle to the principle of freedom to provide services*** *(see, in particular, Kohll, paragraph 41; Smits and Peerbooms, paragraph 72, and Case C-444/05 Stamatelaki [2007] ECR I-3185, paragraph 30)." [...]*

"In those circumstances, national legislation such as that at issue in the main proceedings, inasmuch as it makes ***affiliation compulsory****, can be* ***justified on an overriding ground of public interest, namely the objective of ensuring the financial equilibrium of a branch of social security****, such compulsory affiliation being a suitable means of securing the attainment of that objective." [...]*

---

[310] EJC case C-350/07 in the matter of Kattner Stahlbau GmbH versus Maschinenbau- und Metall-Berufsgenossenschaft, judgment of 5 March 2009, marginal nos. 68, 85, 88 and 90. Emphasis by the author. In addition, the special concerns of the workmen's compensation insurance companies are taken into account in the new Directive 2009/138. Recital 86 explains that national legislation that was specifically issued for workmen's compensation insurance takes precedence. Art. 207, which will enter into force on 1 November 2012, repeats this proviso. Specific national provisions on workmen's compensation insurance will therefore take precedence over FOE and FMS in future as well. Should, therefore, a member state provide for a legal monopoly for workmen's compensation insurance, this must also, and especially, be respected under Directive 2009/138. Regarding the exemption of statutory social security systems, cf. footnotes 312 and 329, regarding general good cf. footnote 313, regarding the new Directive 2009/138 in general cf. footnote 297.

[311] Corresponding to art. 101 and 102 TFEU.

"Moreover, as regards the scope of cover such as that provided by that statutory insurance scheme, it is conceivable that, as MMB submits, if compulsory affiliation were to be applied only in respect of certain benefits, such as those aimed at prevention, as Kattner surmises in its submissions, undertakings employing, for example, ***young employees in good health*** *engaged in non-dangerous activities would seek* ***more advantageous insurance terms*** *from* ***private insurers****. The progressive* ***departure*** *of those* ***'good' risks*** *would leave* ***employers' liability insurance associations*** *such as MMB with an* ***increasing share of 'bad' risks****, thereby* ***increasing the cost of benefits****, particularly for undertakings with* ***older employees engaged in dangerous activities; those associations could no longer offer pensions at an acceptable cost to such undertakings****. Such a situation would arise particularly in a case where, as in the main proceedings, the statutory insurance scheme at issue, inasmuch as it applies the principle of solidarity, is characterised, in particular, by the absence of a strictly proportionate link between contributions and risks insured (see, by analogy, Albany, paragraphs 108 and 109)."*

The following legal situation arises from the judgment:

- An insurance organisation that is based on the principle of solidarity and that performs a social function, does not represent an undertaking within the meaning of EU competition law, which is why EU competition law does not apply (see also p. 121 et seq. below).
- The considerable threat to the financial equilibrium of a social security system represents an overriding public interest which justifies a restriction of FMS.
- A statutory insurance or a corresponding legal monopoly is justified by this overriding public interest or, to put it in a positive way, by the objective of guaranteeing the financial equilibrium of the social security system.
- The discontinuation of the monopoly leads to the destruction of the solidarity and of the financial equilibrium of the statutory insurance which could then no longer guarantee insurance coverage for all at affordable premiums. Hence, the social function can no longer be performed.
- The legal monopoly of statutory insurance is thus a permissible restriction of the FMS.

C. Statutory insurance against natural hazards as an expression of overriding public interest

The KGV statutory insurance monopoly was qualified by the FC, for socio-political and police reasons, as a justified restriction of economic freedom (see p. 90 et seq. above). Even within the framework of the EU and even more so in the face of climate change, a statutory insurance against natural hazards would be recognised as a social function. If nothing else, this is documented in the new Treaty of Lisbon. In the Third Part of the TFEU, which contains *"Union policies and internal actions"*, a new title XXIII, *"Civil Protection"*, was introduced. It states, among other things, that *systems for preventing and protecting against natural disasters shall be encouraged* (art. 196 par. 1 TFEU). Furthermore, *the EU shall aim to support and complement Member States' action at national, regional and local level in risk prevention and in responding to natural disasters* (art. 196 par. 1 subpar. 2(a) TFEU; see also p. 138 et seq. below). The fact that comprehensive insurance against natural hazards is dependent on a legal monopoly in order to guarantee solidarity, financial equilibrium and accessibility for all has been demonstrated on numerous occasions already. As with all other insurance providers that have a social function, the elimination of the monopoly would destroy the financial equilibrium of the system and thus make it impossible for them to perform their function. Statutory insurance against natural hazards is thus an expression of overriding public interest as well and justifies the restriction of FMS (with respect to competition law, see p. 121 et seq. below).

Hence, this represents another foundation for EU compatibility of the statutory insurance system against natural hazards.

D. No exclusion of overriding public interest by harmonisation

It should be pointed out in this context that the elimination of the German building insurance monopolies by Directive 92/49 (see p. 111 above) does not, for instance, entail the exclusion of the general good described above due to harmonisation.

To begin with, it should be stated that damage prevention and damage abatement were not integrated in German building insurances. They cannot therefore be compared with the KGV. Therefore, with the elimination of the monopoly by Directive 92/49 no statement can be made regarding monopolies such as that of the KGV. These are not a subject matter for Directive 92/49 in substance.

In addition, as demonstrated above, the general good is not defined conclusively, but this term is regularly filled by the ECJ with new content according to the change in values over time. Only in recent years, long after Directive 92/49 had taken effect, did the ECJ recognise the considerable threat to the financial equilibrium of a social security system as a public interest. This sector is not harmonised by Directive 92/49. Apart from that, it falls within the competence of the Member States to determine what constitutes a social system and how it is organised[312]. In addition, even the

312 Although the EU and the Member States share the responsibility in social politics, this only applies to the aspects referred to in the TFEU (art. 4 par. 2(b) TFEU). These relate only to the social protection of the employee, where the EU merely has the authority to support and to promote the measures of the Member States (cf. art. 151 et seq. TFEU). All other powers lie exclusively within the competence of the Member States in accordance with the principle of conferral of competences (cf. art. 5

introduction of the new EU objectives to promote systems for preventing and protecting against natural disasters shows that this sector is still in the very early stages; this is the first time that climate change as a global problem is reflected in the TFEU. Furthermore, it has been determined that precisely no harmonisation should take place in this regard. Art. 196 par. 2 TFEU reads: *"The European Parliament and the Council, acting in accordance with the ordinary legislative procedure, shall establish the measures necessary to help achieve the objectives referred to in paragraph 1, **excluding any harmonisation** of laws and regulations of the Member States."* Every Member State should decide for itself which system is appropriate for it to protect against natural disasters.

Finally, Directive 92/49 itself is open to restrictions based on the newly recognised meaning of the term overriding public interest. In its 19th recital, it contains a general provision in favour of the general good. Accordingly, each Member State must grant unhindered access to the market unless legal provisions that protect the general good restrict market access[313]. Thus, if overriding public interest is present, as is the case here, the scope of application of Directive 92/49 can be restricted.

Hence, no harmonisation at all has taken place as yet with respect to statutory insurance against natural hazards. Instead, the Member States have complete freedom here regarding the configuration and can readily provide for a statutory monopoly if they consider it appropriate. They can invoke overriding public interest/the general good in this regard.

TEU). In addition, it should be pointed out that Directive 92/49 does not apply from the outset to "insurance forming part of a statutory system of social security" (art. 2 par. 2 Directive 92/49 in conjunction with art. 2(d) Directive 73/239). This also applies under the new Directive 2009/138. Art 3 of the latter, which will enter into force on 1 November 2012, excludes *"Statutory social security schemes"* from the scope of application of Directive 2009/138 without change. Cf. also footnotes 297–299 on the above-mentioned Directive.

313 Art. 28 Directive 92/49 repeats the restriction as a principal provision at the expense of the FMS and FOE, art. 32 par. 4 at the expense of the FOE. In the newer Directive 2009/138, national legislation that safeguards the general good, that is to say, they enforce public interests, are reserved in any case. The FOE and the FMS are accordingly restricted as before. Art. 180, *"General good"*, which will enter into force on 1 November 2012, states:
"Neither the Member State in which a risk is situated nor the Member State of the commitment shall prevent a policy holder from concluding a contract with an insurance undertaking authorised under the conditions of Article 14 as long as that conclusion of contract does not conflict with legal provisions protecting the general good in the Member State in which the risk is situated or in the Member State of the commitment."
Hence, the legal situation will remain unchanged in future as well, i.e. the general good will continue to take precedence over FOE and FMS. Cf. also footnotes 310, 329 and 361 as well as footnote 297 on Directive 2009/138 in general.

6. Statutory insurance as a provision of general interest services and as a service of general (economic) interest

A. Terms

The various terms were created, developed and consolidated by the ECJ, the EU Member States and the EU Commission in the course of the case law on art. 106 par. 2 TFEU [86 par. 2] and the Amsterdam, Nice and Lisbon treaty revisions. Today, the following understanding of concepts can be taken as a starting point:

- *Services of general interest or facilities for the provision of general interest services:* this refers to all economic and non-economic services provided by state authorities in the interest of the general public and are therefore interlinked by them with specific obligations relating to the common good[314]. This is an umbrella term. The term "facility for the provision of general interest services" is used as a synonym for this[315]. The term "service of general interest" can be found in protocol no. 26 "on services of general interest" to the Treaty of Lisbon and in the General Services Directive 2006/123 (see p. 120 et seq., 124 et seq. and 130 et seq. below). Facilities and services from a wide variety of sectors are subsumed under this pair of terms, e.g. energy and water supply, traffic undertakings, the postal service, certain insurance organisations and benefits, health care, cultural and educational institutions or radio broadcasting.

- *Public services:* this term refers, on the one hand, to a facility that provides services with obligations to serve the common good and, on the other hand, to the service rendered under the obligation to serve the common good itself. It is used as a synonym both for facilities for the provision of general interest services as well as for services of general interest[316]. The term is not used in the TEU and the TFEU (nor up until now in the TEU and the TEC), but is used in the General Service Directive 2006/123.

- *Services of general economic interest:* this refers to economic services that are associated with obligations to serve the common good, i.e. services that are rendered in the interest of the community[317]. These services are in principle subject to internal market and competition rules which will then, however, have to stand down in favour of the obligations to serve the common good. The term is used on numerous occasions in the legal bases, in particular in the EU Fundamental Rights Charter (art. 36), art. 14 and 106 TFEU [16 and 86], protocol no. 26 "on

---

314 COM communication Services of General Interest, p. 4.

315 COM communication Services of General Interest in Europe, p. 4 et seq., annex II.

316 Kallmayer/Jung, in: Calliess/Ruffert, eGV 16 N2; Voet van Vormizeele, in: Schwarze, EGV 86 N 53; COM communication Services of General Interest in Europe, annex II.

317 COM communication Services of General Interest, p. 4 et seq.; White Paper COM Services of General Interest, p. 27.

services of general interest" to the Treaty of Lisbon and in the General Service Directive 2006/123. These include, for example, services from the energy and water supply sectors, postal service, telecommunication, traffic or waste management.

- *Non-economic services of general interest:* these are non-economic services. The term is used in protocol no. 26 *"On services of general interest"* to the Treaty of Lisbon and in the General Services Directive 2006/123. These services are not subject either to the internal market nor competition rules. Social security systems in particular are subsumed under this term[318]. It must, however, be emphasised that an across-the-board distinction between economic services and non-economic services does not exist. That depends on the configuration on a case-by-case basis which may differ from one Member State to another. A service that is the same in itself may be an economic service of general interest in one Member State while it is a non-economic service in another State[319]. The Member States have considerable discretionary powers in this regard (see p. 120 et seq., p. 125 et seq. below).
- *Universal service:* this term was coined most notably in the discussion on the obligations to serve the common good in the postal service, energy supply, telecommunication and railway sectors. It is, however, not restricted to specific sectors but is open as regards content in this respect and can also refer to non-economic services[320]. The term also does not outline a specific scope of services but rather the requirements of the provision of services. Universal services must be provided in such a manner as to be spatially inclusive, accessible to all, of good quality, continuous and especially at a price that is affordable for all. The term universal service is increasingly absorbed in the term general interest services, and services of general interest, and the corresponding services must also meet the above-mentioned requirements.[321]. The term universal service corresponds to the term "basic service provision" (Leistung der Grundversorgung) according to art. 43a par. 4 BV (see p. 97 et seq. above).

B. From Amsterdam and Nice to Lisbon—strengthening of services of general interest

The status of services of general interest has been strengthened considerably in the last ten years. Following a wave of liberalisations and privatisations, a change in

---

318 COM communication Services of General Interest, p. 5.

319 COM communication Services of General Interest, p. 5 et seq.

320 Cf. General Service Directive 2006/123, recital 28, Universal Service Social Services.

321 COM communication Services of General Interest in Europe, annex II; COM communication Services of General Interest, p. 7; cf. Sandmann, p. 54 et seq.

thinking has taken place in favour of services rendered under the obligation to serve the common good, which is reflected in numerous pieces of legislation[322]:

- Services of general economic interest have in addition found their way into the Union's *"Principles"* section in the Treaty of Amsterdam[323] (art. 16 TEC, now art. 14 TFEU), having previously been incorporated as a provision only in the chapter on TEC competition rules (art. 86 TEC, now art. 106 TFEU). This has turned them from an exemption clause into an independent legally protected interest and objective of the Union.
- At the same time, art. 3 par. 1(g) TEC, which contained the *system of undistorted competition* as one of the objectives of the EU, has been *deleted* in the course of Lisbon in response to French pressure. Within the framework of the TEU, there is talk of a *"highly competitive social market economy, aiming at full employment and social progress"* (art. 3 par. 3 TEU). In addition *"open"* was replaced by *"social"* market economy (cf. the former art. 4 par. 1 TEC). Thus, while services of general interest are still contained in the *"Principles"* of the Union, *"competition"* has dropped out. The fact that the "principles" must be taken into account in interpreting the provisions of the TFEU will have legal consequences. To wit, art. 3 par. 1(g) TEC has so far been regarded a guarantee of existence for the institution of "competition" by the ECJ. Through the omission of the provision mentioned, this guarantee has been weakened and the deletion is by no means merely of symbolic nature, as contended in some cases[324].
- Through the Treaty of Lisbon, by means of protocol no. 26 *"On services of general interest"*[325], the significance and safeguarding of services of general interest was further increased by and large. In particular, the autonomy of the Member States was strengthened in this sector (for wording, see p. 124 et seq. below). The protocol has the same legal obligation as the TFEU (art. 51 TEU), therefore it represents primary law and thus takes precedence over the Directive (secondary law, see footnote 348 below).

---

322 JUNG, in: CALLIESS/RUFFERT, EGV 86 N 41; KALLMAYER/JUNG, in: CALLIESS/RUFFERT, EGV 16 N 17 et seq.; Hatje, in: SCHWARZE, EGV 16 N 1; VOET VAN VORMIZEELE, in: SCHWARZE, EGV 86 N 2, 53.

323 Treaty of Amsterdam amending the Treaty on the European Union, the Treaties establishing the European Communities and certain related acts, dated 2 October 1997, OJEU C340 of 10 November 1997, p. 1 et seq., consolidated version p. 173 et seq. The Treaty of Amsterdam entered into force on 1 May 1999.

324 HATJE, in: SCHWARZE, EGV 3 N 14; Brinker, in: SCHWARZE, EGV 81 N 41. The fact that it is now stated in protocol no. 27 to the Treaty of Lisbon *"on the internal market and competition"* that *"the internal market as set out in article 3 of the Treaty on European Union includes a system ensuring that competition is not distorted"* has no effect on this either. Competition is precisely no longer an objective of the EU but instead "merely" an element of the internal market, among other things, that is designed to serve the objective of the social market economy which is geared towards social progress.

325 OJEU C306 of 17 December 2007, p. 158 et seq.

- The Charter of Fundamental Rights of the European Union ("CFR")[326] is now legally binding in accordance with the Treaty of Lisbon (art. 6 TEU) so that even art. 36 CFR, which emphasises the autonomy of the Member States with respect to general economic interest services, becomes legally binding and thus more significant.
- The Member States were strengthened with regard to the division of competences and the principle of subsidiarity through the Treaty of Lisbon. On the one hand, the principle of conferral of competences of the EU has been introduced. According to this, the EU has competence only if it has a clear jurisdiction in the TEU or in the TFEU; otherwise, the Member States have competence. On the other hand, the EU may *not* operate, in accordance with the more broadly worded principle of subsidiarity, in sectors where it does not have exclusive competence, if the objectives are sufficiently achieved through *regional* and *local* regulations of the Member States (art. 5 TEU, previously art. 5 TEC; see p. 124 et seq. below).
- In the General Service Directive 2006/123[327] which entered into force on 28 December 2006 and which had to be implemented by 28 December 2009, numerous provisions are contained in favour of general (economic) interest services, the general good, universal service and overall in favour of the obligations to serve the common good (see p. 130 et seq. below).

C. Statutory insurance as an undertaking within the meaning of EU competition law?

EU competition law applies only to "undertakings" (cf. art. 101 and 102 TFEU [81/82]). In this regard, it takes a broad, that is to say, functional, notion of undertaking. According to this, any entity that performs an economic activity, irrespective of its legal structure and funding, is an undertaking. Any activity that consists of offering certain goods or services on a market is considered an economic activity. The intention of making a profit is *not* required. Any corresponding activity, even if it is carried out by an entity regulated by public law, even by an administrative department, is included[328].

Here, however, it is appropriate to come back to the ruling in Kattner/MMB (see p. 113 et seq. above), which is one in a long list of comparable rulings. In spite of the

326 Charter of Fundamental Rights of the European Union dated 12 December 2007, OJEU C303 of 14 December 2007, p. 1 et seq.

327 Directive 2006/123/EC of the European Parliament and of the Council of 12 December 2006 on services in the internal market, OJEU L376 of 27 December 2006, p. 36 et seq.

328 VOET VAN VORMIZEELE, in: SCHWARZE, EGV 86 N 11, with numerous references to case law; see BRINER, in: SCHWARZE, EGV 81 N 27 regarding the intention of making a profit.

very broad definition of the term undertaking, according to ECJ established case law, an institution that operates in the sector of social security is *not an undertaking.* In recent years, the ECJ has primarily had to deal with health insurances in this regard and stated in each case that the health insurance concerned was acting exclusively in fulfilment of its public *service task, according to the principle of solidarity.* It was exercising a purely social responsibility based on the *principle of solidarity.* It was compelled by law to offer essentially *identical obligatory benefits*, regardless of the amount of contributions. Furthermore, health insurances were *joined together* in a type of *community founded on the basis of solidarity which enables an equalisation of costs and risks between them*[329].

Ultimately, a *non-economic* service of general interest exists in such a case. Which services come under this category is *in the sole power of decision and competence of the Member States* based on protocol no. 26 to the Treaty of Lisbon, art. 2. The autonomy of the Member States is even more comprehensive here than it is for services of general economic interest (see p. 126 et seq. below).

Consequently, such an institution drops out of the scope of application of the EU rules on competition and is thus compatible with EU law.

The abstract definition of an institution in the sector of social security by the ECJ can be transferred to the KGV. Even the KGV statutory insurance against natural hazards is based on the principle of solidarity. Equalisation between good and bad risks takes place necessarily. Premium differentiation is restricted to a narrow range defined by law and the benefit is the same, namely, the replacement value of the building. Even the KGV equalise risks among themselves, in particular in the event of natural disasters. According to ECJ case law, even statutory insurance companies such as those of the KGV thus represent institutions operating in the sector of social security and are therefore exempt from EU competition law.

Hence, another basis for the compatibility of the KGV statutory insurance system with EU law arises from the prerequisites for the application of EU competition law alone.

The same consequence arises if an institution exercises official authority. According to established case law of the ECJ, the undertaking characteristic is then absent as well. It has already been shown that statutory insurance against natural hazards is inevitably associated with the exercise of official authority (see p. 108 et seq. above).

---

[329] Brinker, in: Schwarze, EGV 81 N 28, with numerous references to case law. Cf. in particular ECJ joint cases C-264/01, C-306/01, C-354/01, C-355/01, in the matter of AOK-Bundesverband, judgment of 16 March 2004, marginal nos 51 et seq. Cf. also the judgments of ECJ case C-238/94 in the matter of Mutuelle, judgment of 26 March 1996 as well as ECJ joint case C-159/91 and C-160/91 in the matter of Poucet, judgment of 17 February 1993. Cf. footnotes 310 and 361 regarding the new Directive 2009/138.

D. The significance of art. 14 and 106 TFEU

a) The wording of the relevant provisions

If one were to assume, however, that statutory insurance against natural hazards represents an undertaking or an economic activity, EU competition law would apply in principle (art. 101 TFEU et seq. [81 et seq.]). This does not, however, mean that an economic activity with an obligation to serve the common good is entirely subject to competition law. Instead, the EU Member States have issued a number of provisions that also exempt economic activities in the general interest from EU competition law as well, namely art. 14 and 106 TFEU [16/86] in particular.

The wording of art. 14 and 106 TFEU as well as of protocol no. 26 to the Treaty of Lisbon, of art. 5 TEU and art. 36 CFR is reproduced below:

TFEU

"Art. 14 [16]

Without prejudice to Article 4 of the Treaty on European Union or to Articles 93, 106 and 107 of this Treaty, and given the place occupied by services of general economic interest in the shared values of the Union as well as their role in promoting social and territorial cohesion, the Union and the Member States, each within their respective powers and within the scope of application of the Treaties, shall take care that such services operate on the basis of principles and conditions, particularly economic and financial conditions, which enable them to fulfil their missions. The European Parliament and the Council, acting by means of regulations in accordance with the ordinary legislative procedure, shall establish these principles and set these conditions without prejudice to the competence of Member States, in compliance with the Treaties, to provide, to commission and to fund such services."

"Art. 106 [86]

(1) In the case of public undertakings and undertakings to which Member States grant special or exclusive rights, Member States shall neither enact nor maintain in force any measure contrary to the rules contained in the Treaties, in particular to those rules provided for in Article 18 and Articles 101 to 109.

(2) Undertakings entrusted with the operation of services of general economic interest or having the character of a revenue-producing monopoly shall be subject to the rules contained in the Treaties, in particular to the rules on competition, in so far as the application of such rules does not obstruct the performance, in law or in fact, of the particular tasks assigned to them. The development of trade must not be affected to such an extent as would be contrary to the interests of the Union.

(3) The Commission shall ensure the application of the provisions of this Article and shall, where necessary, address appropriate directives or decisions to Member States"

"Protocol (no. 26)

On services of general interest

THE HIGH CONTRACTING PARTIES,

WISHING to emphasise the importance of services of general interest,

HAVE AGREED UPON the following interpretative provisions, which shall be annexed to the Treaty on European Union and to the Treaty on the Functioning of the European Union:

Article 1

The shared values of the Union in respect of services of general economic interest within the meaning of Article 14 of the Treaty on the Functioning of the European Union include in particular:

- *the essential role and the wide discretion of national, regional and local authorities in providing, commissioning and organising services of general economic interest as closely as possible to the needs of the users;*
- *the diversity between various services of general economic interest and the differences in the needs and preferences of users that may result from different geographical, social or cultural situations;*
- *a high level of quality, safety and affordability, equal treatment and the promotion of universal access and of user rights.*

Article 2

The provisions of the Treaties do not affect in any way the competence of Member States to provide, commission and organise non-economic services of general interest."

TEU

"Art. 5 [5 TFEC]

1 The limits of Union competences are governed by the principle of conferral. The use of Union competences is governed by the principles of subsidiarity and proportionality.

2 Under the principle of conferral, the Union shall act only within the limits of the competences conferred upon it by the Member States in the Treaties to attain the objectives set out therein. Competences not conferred upon the Union in the Treaties remain within the Member States.

3 Under the principle of subsidiarity, in areas which do not fall within its exclusive competence, the Union shall act only if and in so far as the objectives of the proposed action cannot be sufficiently achieved by the Member States, either at central level or at regional and local level, but can rather, by reason of the scale or effects of the proposed action, be better achieved at Union level.

The institutions of the Union shall apply the principle of subsidiarity as laid down in the Protocol on the application of the principles of subsidiarity and proportionality. National Parliaments ensure compliance with the principle of subsidiarity in accordance with the procedure set out in that Protocol.

4 Under the principle of proportionality, the content and form of Union action shall not exceed what is necessary to achieve the objectives of the Treaties.

The institutions of the Union shall apply the principle of proportionality as laid down in the Protocol on the application of the principles of subsidiarity and proportionality."

CFR

"Art. 36

Access to services of general economic interest

The Union recognises and respects access to services of general economic interest as provided for in national laws and practices, in accordance with the Treaties, in order to promote the social and territorial cohesion of the Union."

b) The significance of art. 14 and 106 TFEU

Art. 14 TFEU [16] has the following significance:

- It establishes the principle that the EU and the Member States are to ensure that the framework conditions for the services of general economic interest are organised such that the corresponding institutions can fulfil their function. Art. 14 protects the economic activity which is under the obligation to serve the common good[330].
- The economy, which is oriented towards serving the common good, is protected by art. 14 as an *independent kind of* economy and is no longer a mere exception from competition law; JUNG speaks of the *"recognition as an independent economic model (translation, original quote in German)."*[331]. This confirms that

[330] HATJE, in: SCHWARZE, EGV 16 N 4.

[331] JUNG, in: CALLIESS/RUFFERT, EGV 86 N 41.

it represents a *"key element of the European social model"* and constitutes an important *component of the European values*[332].

- Since it is a basic provision, it must be taken into account in the interpretation of all other provisions, in particular art. 106 TFEU[333].
- It is within the competence of the Member States to determine which services are considered "services of general economic interest" and are therefore privileged with respect to competition law. This term should be interpreted broadly and the EU institutions can intervene only in the event of blatant misallocations, thus they merely execute a control function in case of misuse[334]. This is also evident from the wording *"without prejudice to the competence of Member States, in compliance with the Treaties, to provide [...] such services."* This part of the sentence had been newly incorporated in art. 14 with the Treaty of Lisbon. In addition, the autonomy of the Member States has been substantially strengthened by protocol no. 26.
- In this context, reference should be made to art. 36 CFR. This provision, which is now legally binding, states that the EU must recognise and respect access to services of general economic interest *"as provided for in national laws and practices [...]"*. This should be considered as further confirmation of the autonomy of the Member States[335]. Art 36 CFR does not give an actionable, individual entitlement to benefits for a service. It does, however, obligate the EU to respect the principle of universal service and should be taken into account for all other EU regulations and in interpreting the Treaty provisions, in particular art. 106 TFEU. In art. 36 CFR, the EU also sets another social objective from which it may no longer deviate in future[336].
- The view that the wording *"Without prejudice to [...] Article 106 [86]"* restricts the scope of art. 14[337] can *no longer* be maintained in the light of the Treaty of Lisbon. Due to the fact that the importance has clearly been shifted overall in favour of general interest services, art. 106 is merely a special provision which implements the principle of art. 14 for the competition sector.

Art. 106 TFEU [86] now enforces the principle of art. 14 TFEU in EU competition law:

---

332 Kallmayer/Jung, in: Calliess/Ruffert, EGV 16 N 10; Communication COM Services of General Interest in Europe, marginal no. 7, 64; White Paper COM Services of General Interest, marginal no. 2.1; Voet van Vormizelle, in: Schwarze, EGV 86 N 53.

333 Hatje, in: Schwarze, EGV 16 N 5; Kallmayer/Jung, in: Calliess/Ruffert, EGV 16 N 9.

334 Kallmayer/Jung, in: Calliess/Ruffert, EGV 16 N 8.

335 Voet van Vormizeele, in: Schwarze, CFR 36 N 5.

336 Voet van Vormizeele, in: Schwarze, CFR 36 N 4 et seq.

337 Hatje, in: Schwarze, EGV 16 N 9.

- *"Exclusive rights"* within the meaning of art. 106 par. 1 TFEU refers in particular to legal monopolies[338]. Considering case law, this is probably the most frequent form of arrangement, intended to make it possible for an undertaking to fulfil its task of serving the common good.
- If a service of general economic interest is provided and if the application of EU competition law, i.e. the abolition of the monopoly, were to result in the fact that the service could no longer be provided for the purpose of the obligation to serve the common good, then EU competition law *in particular*[339] would not apply according to art. 106 par. 2 TFEU, and the legal monopoly can be maintained[340]. VOET VAN VORMIZEELE refers to a *"general public service clause (translation, original quote in German)"*[341].
- The Member States decide on what they consider to be a service of general economic interest. They are effectively autonomous in this regard. This is confirmed once again by protocol no. 26, art. 1, and also arises as a result of the principle of subsidiarity and the division of competences between the EU and the Member States. It is not only the central or federal discretionary powers, but also the discretionary powers at the regional and local level that must be respected. The different practices and needs of the Member States, due not least to geographical, social and cultural differences, have priority. The future will show whether, in the light of these protocols, the EU will even be entitled to any control function in case of misuse[342] at all (see above regarding art. 14 TFEU and art. 36 CFR).
- The application of competition law must not mean that fulfilment of the task becomes impossible. Suffice that the provision of services by the institution under economically sustainable conditions is *compromised*[343].
- This is where universal service enters into it. Its obligation to provide services to all, comprehensively, continuously, at good quality and at a price that is affordable for all, was transferred to services of general economic interest (see protocol no. 26, art. 1, last bullet point, and art. 36 CFR). If it *is no longer possible* for the

338 VOET VAN VORMIZEELE, in: SCHWARZE, EGV 86 N 24; JUNG, in: CALLIESS/RUFFERT, EGV 86 N 16.

339 According to art. 106 par. 2 TFEU, *"the rules contained in the Treaties, in particular (...) the rules on competition"*, do not apply. This means that, if the prerequisites of art. 106 par. 2 TFEU exist, even the fundamental freedoms of the TFEU such as, for example, the FOE and the FMS, do not apply. Hence, an analysis according to art. 106 par. 2 TFEU is sufficient per se, and an additional evaluation as to whether the facility for the provision of general interest services would also entail the non-application of fundamental freedoms, is obsolete. Cf. VOET VAN VORMIZEELE, in: SCHWARZE, EGV 86 N 52.

340 JUNG, in: CALLIESS/RUFFERT, EGV 86 N 34 et seq.; VOET VAN VORMIZEELE, in: SCHWARZE, EGV 86 N 50 et seq.

341 VOET VAN VORMIZEELE, in: SCHWARZE, EGV 86 N 53.

342 VOET VAN VORMIZEELE, in: SCHWARZE, EGV 86 N 62.

343 JUNG, in: CALLIESS/RUFFERT, EGV 86 N 45; VOET VAN VORMIZEELE, in: SCHWARZE, EGV 86 N 69.

institution (were it to be exposed to competition) to provide the service within the framework of the above-mentioned obligation, then the legal monopoly can remain in force[344].

- It is exactly in this way that the institution should be allowed to finance the provision of a service that, in some cases, is in deficit by a cost-efficient provision of service, that is to say, to undertake cross-subsidisation within the monopoly. With respect to insurance, this is designed to allow a balance of good and bad risks. The discontinuation of the monopoly would lead to "cherry-picking" by the private competition and would throw the institution off balance. This would compromise its ability to fulfil its task, which is exactly what is to be prevented[345].
- The fact that the non-application of the Treaty provisions, in particular provisions on competition, is necessary constitutes the final prerequisite. This involves the examination of proportionality. The exemption from competition rules must be appropriate, necessary and proportionate[346]. However, the examination is rather formal in character and must not undermine the decision of a Member State to provide a service of general interest. The ECJ usually examines whether the exemption is required in order to fulfil the task under economically sustainable conditions[347]. This is, however, consistently the case if the Member State has decided to perform a task within the meaning of universal service.
- If the conditions of art. 106 par. 2 TFEU are fulfilled, then even EU secondary law[348], in particular the directives, must be waived[349].
- Until now, a restrictive interpretation or application of the previous provision of art. 86 par. 2 TEC had been assumed[350]. This approach can no longer be maintained for the new art. 106 par. 2 TFEU in the light of the numerous changes by Lisbon since, as stated above, a clear shift has taken place in favour of an economy that is oriented towards serving the common good.

---

344 Voet van Vormizeele, in: Schwarze, EGV 86 N 69; Communication COM Services of General Interest, p. 11.

345 Voet van Vormizeele, in: Schwarze, EGV 86 N 69.

346 Voet van Vormizeele, in: Schwarze, EGV 86 N 68.

347 Voet van Vormizeele, in: Schwarze, EGV 86 N 69, with references to case law.

348 EU secondary law must be distinguished from EU primary law. EU primary law is comparable with constitutional law and comprises primarily the Treaties of the EU, that is to say, the constituting Treaties, i.e. currently the TEU and the TFEU. EU primary law forms the foundation, the framework and the limit of EU secondary law. The latter is comparable with ordinary legislation and is derived from primary law. This includes in particular EU regulations and directives. EU primary law is above EU secondary law in the hierarchy and thus takes precedence over the latter. If a directive contradicts EU primary law, it must be waived. Oppermann/Classen/Nettesheim, p. 164 et seq., 173 et seq.

349 Voet van Vormizeele, in: Schwarze, EGV 86 N 54.

350 Jung, in: Calliess/Ruffert, EGV 86 N 35.

These principles and correlations have already been recognised in earlier case law, albeit only for the purpose of an exception. Two judgments, which are still being quoted even today by the EU Commission as landmark rulings, will be commented on briefly below[351]:

- In the much-noted "Corbeau" judgment, the ECJ commented on the Belgian postal service monopoly[352]. It first held that the postal service indisputably provides services of general economic interest, namely the delivery of mail in the entire sovereign territory to the benefit of all inhabitants, at standard charges, of consistent quality and regardless of special cases and of profitability. Based on this, the ECJ concluded that the postal service must have the possibility of balancing profitable and non-profitable areas of activity. It further held that private entities are not permitted to provide any services from the postal services' area of monopoly. For, according to the ECJ, this would allow private entities to focus on the profitable sectors whereby the economic equilibrium of the postal services would be disrupted and the postal services would ultimately be hindered in fulfilling its tasks.
- In the "Almelo" judgment, the ECJ considered the electricity market in the Netherlands[353]. It specifically had to hear the dispute between a municipality and a local energy supply company, on the one hand, and a regional energy supply company on the other hand. This involved the question of to what extent an exclusive obligation to take delivery on the part of the municipalities compared to the regional energy supply company, or an import ban for municipalities is permissible, and whether they were permitted to obtain energy or to import it from third parties themselves. In considering this question, the ECJ first of all held that the regional energy supply company was providing a service of general economic interest. This is substantiated by the fact that

  "such an undertaking had to ensure the uninterrupted supply of electricity in the entire concession territory for all customers, [...], at the quantities requested at any time, at standard rates and under conditions that may differ only according to objective criteria and that apply to all customers."

  The costs thereby borne by the regional energy supply company, as well as the various regulations (in particular environmental regulations) that it had to observe were taken into account as well. The ECJ held that the exclusion of other economic participants from competition is permitted insofar as this was required

---

351 Communication COM Services of General Interest, p. 6 footnote 8. Cf. also the numerous references to case law by Voet van Vormizeele, in: Schwarze, EGV 86 N 65, 69.

352 ECJ case C-320/91 in the matter of Corbeau, judgment of 19 May 1993.

353 ECJ case C-393/92 in the matter of Almelo, judgment of 27 April 1994.

to allow the regional energy supply company to fulfil its tasks. In its case law, the ECJ expressly referred to the "Corbeau" judgment.

E. Statutory insurance against natural hazards as a facility for the provision of general interest services and as a service of general economic interest

Statutory insurance such as that of the KGV represents a facility for the provision of general intererst services in any case. The decision on whether insurance against natural hazards should be formulated as a facility for the provision of general interest services, or provided for the purpose of a universal service would effectively be up to Switzerland alone, if Switzerland were an EU Member State. It would have considerable autonomy in this regard.

On the one hand, statutory insurance may qualify as a non-economic service or a task for the purpose of a facility in the social security sector (see p. 112 et seq., 121 et seq. above). Thus, it falls outside the scope of application of EU competition law.

On the other hand, the statutory insurance system can also qualify as a service of general economic interest. If the legal monopoly were to be abolished, statutory insurance against natural hazards could no longer be guaranteed under economically sustainable conditions. Solidarity among policyholders (cross-subsidisation between good and bad risks) could no longer be maintained. The financial equilibrium would be thrown out of joint and the KGV could no longer fulfil their task. The conditions of art. 106 par. 2 TFEU are therefore met, and for this reason the legal monopoly can continue to remain in force.

F. Conclusion—no application of EU competition rules

Ultimately, it does not make a difference what type of facility for the provision of general interest services, and what type of service of general interest, is present. What matters is that a statutory insurance system such as that of the KGV represents such a facility and therefore the EU rules on competition do not apply.

The statutory insurance system presented in the example of the KGV, in its current form, *with a legal monopoly*, is thus compatible even with EU competition law.

G. No application of the General Service Directive

The General Service Directive 2006/123[354] is yet further evidence of how strongly the emphasis has shifted in favour of economies that are under the obligation to serve the common good.

[354] For reference, see footnote 327.

The proposal for the Directive by the EU Commission of 25 February 2004[355] was amended in numerous points by the European Parliament and was weakened considerably in favour of general interest services. The EU Commission's liberalisation efforts went much too far for the Member States. The Directive was finally approved on the basis of the new draft by the EU Parliament[356].

The key provisions of the above-mentioned Directive are reproduced below. The following essentially arises as a result:

- Overriding public interest is reserved or takes priority over the FMS.
- The same applies to services associated with the exercise of official authority.
- Non-economic services of general interest are reserved or do not come under the scope of application of the Directive.
- Although, in principle, services of general *economic* interest come under the scope of application of the Directive, neither the liberalisation nor the privatisation of corresponding services and institutions can be required on this basis. Service monopolies, universal service obligations and relevant aids must not be abolished and take priority over the Directive. Furthermore, the Directive does not restrict the Member States' right to determine which deliverables are considered services of general economic interest. As a result, and substantively, the Directive therefore does not apply to services of general economic interest either.
- The Directive does not apply to the tax sector. This is of particular importance for the regulation of the Consorcio. It thereby ensures that the special regulation of the Consorcio is not affected by the Directive in EU law.

From this, it follows that statutory insurance against natural hazards such as that provided by the KGV is on no account affected by the General Service Directive. Since a facility for the provision of general interest services is involved in each case, either of economic or of non-economic nature, the Directive either does not apply from the outset, or else it must not interfere in the configuration, in particular the legal monopoly. A statutory insurance system according to the example of the KGV is thus consistent with the General Service Directive as well.

---

355 COM(2004) 2 final/2.

356 Legislative resolution by the European Parliament dated 16 February 2006 on the proposal for a Directive by the European Parliament and the Council on services in the internal market COM(2004)0002-C5-0069/2004-2004/0001 (COD); NZZ of 17 February 2006, p. 21, "EU-Parlament schwächt Dienstleistungs-Richtlinie ab" (EU Parliament weakens Service Directive); Amended proposal by the COM for a Directive by the European Parliament and the Council on services in the internal market dated 4 April 2006, COM(2006) 0160 final; NZZ of 5 April 2006, "EU-Dienstleistungs-Richtlinie auf Kurs gebracht—Kommision folgt Parlamentsvorschlag" (EU Service Directive back on track—Commission follows Parliament proposal).

The key provisions of the General Service Directive are reproduced below:

"Recitals

(8) It is appropriate that the provisions of this Directive concerning the freedom of establishment and the free movement of services should apply only to the extent that the activities in question are open to competition, so that they do not oblige Member States either to liberalise services of general economic interest or to privatise public entities which provide such services or to abolish existing monopolies for other activities or certain distribution services. [...]

(15) This Directive respects the exercise of fundamental rights applicable in the Member States and as recognised in the Charter of fundamental Rights of the European Union and the accompanying explanations, reconciling them with the fundamental freedoms laid down in Articles 43 and 49 of the Treaty. [...]

(17) This Directive covers only services which are performed for an economic consideration. Services of general interest are not covered by the definition in Article 50 of the Treaty and therefore do not fall within the scope of this Directive. Services of general economic interest are services that are performed for an economic consideration and therefore do fall within the scope of this Directive. However, certain services of general economic interest, such as those that may exist in the field of transport, are excluded from the scope of this Directive and certain other services of general economic interest, for example, those that may exist in the area of postal services, are the subject of a derogation from the provision on the freedom to provide services set out in this Directive. This Directive does not deal with the funding of services of general economic interest and does not apply to systems of aids granted by Member States, in particular in the social field, in accordance with Community rules on competition. This Directive does not deal with the follow-up to the Commission White Paper on Services of General Interest. [...]

(28) This Directive does not deal with the funding of, or the system of aids linked to, social services. Nor does it affect the criteria or conditions set by Member States to ensure that social services effectively carry out a function to the benefit of the public interest and social cohesion. In addition, this Directive should not affect the principle of universal service in Member States' social services. [...]

(40) The concept of 'overriding reasons relating to the public interest' to which reference is made in certain provisions of this Directive has been developed by the Court of Justice in its case law in relation to Articles 43 and 49 of the Treaty and may continue to evolve. The notion as recognised in the case law of the Court of Justice covers at least the following grounds: public policy, public security and public health, within the meaning of Articles 46 and 55 of the Treaty; the maintenance of order in society; social policy objectives; the protection of the recipients of services;

consumer protection; the protection of workers, including the social protection of workers; animal welfare; the preservation of the financial balance of the social security system; the prevention of fraud; the prevention of unfair competition; the protection of the environment and the urban environment, including town and country planning; the protection of creditors; safeguarding the sound administration of justice; road safety; the protection of intellectual property; cultural policy objectives, including safeguarding the freedom of expression of various elements, in particular social, cultural, religious and philosophical values of society; the need to ensure a high level of education, the maintenance of press diversity and the promotion of the national language; the preservation of national historical and artistic heritage; and veterinary policy. [...]

(71) The mutual evaluation process provided for in this Directive should not affect the freedom of Member States to set in their legislation a high level of protection of the public interest, in particular in relation to social policy objectives. Furthermore, it is necessary that the mutual evaluation process take fully into account the specificity of services of general economic interest and of the particular tasks assigned to them. This may justify certain restrictions on the freedom of establishment, in particular where such restrictions pursue the protection of public health and social policy objectives and where they satisfy the conditions set out in Article 15(3)(a), (b) and (c). [...]

Chapter 1

General provisions

Article 1 Subject matter

[...]

(2) This Directive does not deal with the liberalisation of services of general economic interest, reserved to public or private entities, nor with the privatisation of public entities providing services.

(3) This Directive does not deal with the abolition of monopolies providing services nor with aids granted by Member States which are covered by Community rules on competition.

This Directive does not affect the freedom of Member States to define, in conformity with Community law, what they consider to be services of general economic interest, how those services should be organised and financed, in compliance with the State aid rules, and what specific obligations they should be subject to.

Article 2 Scope

[...]

(2) This Directive shall not apply to the following activities:

(a) non-economic services of general interest;

[...]

(i) activities which are connected with the exercise of official authority as set out in Article 45 of the Treaty;

[...]

(3) This Directive shall not apply to the field of taxation."

7. Statutory insurance against natural hazards as a permissible compensation scheme according to the EU law governing State aid

A. Configuration as a facility for the provision of general interest services as state aid?

In recent years, services of general economic interest have become an issue in the state aid sector as well, either because Member States have reported corresponding schemes to the EU Commission as (permissible) "aid" and/or such schemes were challenged in court as prohibited state aid. Non-economic facilities for the provision of general interest services, however, are excluded from the scope of EU competition law a priori and thus cannot infringe the prohibition on aid.

The definitive provision in the TFEU reads:

"Aids granted by States

Article 107 [87]

(1) Save as otherwise provided in the Treaties, any aid granted by a Member State or through State resources in any form whatsoever which distorts or threatens to distort competition by favouring certain undertakings or the production of certain goods shall, in so far as it affects trade between Member States, be incompatible with the internal market.

[...]"

State aid is prohibited in principle. The concept of aid is relatively broad. Any voluntary grant of any kind of economic advantage without a corresponding consideration, any pecuniary advantage, comes under this in principle. Thus, the concept extends far beyond what is normally thought of as subsidies[357].

[357] Bär-Bouyssière, in: Schwarze, EGV 87 N 22, 27, 30.

Especially in the sector of services of general economic interest, the ECFI[358] and the ECJ have developed case law on the legitimacy of aid. According to this, grants in the broadest sense are *not* aid if they *"are to be considered compensation for services provided by the recipient [facility for the provision of general interest services] to comply with non-profit-making obligations"*[359]. The term "non-profit-making" simply means "oriented towards serving the common good". Four prerequisites must be met in this regard. For more details, we refer to the case law[360].

B. Balancing of risks in a solidary insurance system does not represent aid

Recently, the ECFI had to deal with a dispute between the British United Provident Association Ltd. (BUPA) and the EU Commission which, following an aid announcement by Ireland regarding the "risk equalisation scheme" ("RES") did not raise any objections and issued a corresponding decision[361].

It involved the introduction of the RES compensation system in the private health insurance sector (health insurance has been privatised in Ireland, there are no longer any state providers). According to this, a benchmark is specified for the average risk profile of all policyholders. Then, the actual risk profile of the policyholders is determined based on their specific insurance portfolio. Those insurers that show a more favourable risk profile compared to the benchmark on this basis must pay compensation payments to those insurance providers that are worse than the benchmark. In other words, insurance providers with too many good risks must pay compensation to insurance providers with too many bad risks. BUPA did not want to submit to this Irish legal provision and appealed against the positive decision of the EU Commission for breach of EU competition law, in particular infringement of the prohibition of State aid.

The ECFI decreed concerning this matter:

---

358 European Court of First Instance. It has jurisdiction for appeals against aid decisions of the EU Commission.

359 Landmark decision ECJ case C-280/00, in the matter of Altmark, judgment of 23 July 2003; ECFI case T-289/03, in the matter of BUPA, judgment dated 12 February 2008, marginal no. 89; Bär-Bouyssière in: Schwarze, EGV 87 N 35.

360 See previous footnote.

361 ECFI case T-289/03, in the matter of BUPA, judgment dated 12 February 2008. The judgment in the matter of BUPA is reflected in the new Directive 2009/138. Thus, in its recitals 84 and 85, the special provisions of national law that are designed to safeguard the social character and general good are reserved for private and/or optional health insurances as well. In recital 85, provisions that stipulate participation in a *"loss compensation system"* are even reserved explicitly. Art. 206 of the above-mentioned Directive which will enter into force on 1 November 2012 repeats the general good proviso and the corresponding national provisions in the sector of private and/or optional health insurance. Cf. also footnote 313 regarding general good, and footnote 297 regarding Directive 2009/138 in general.

"41. However, the Commission concluded that the compensation provided by the RES none the less did ***not constitute State aid within the meaning of Art. 87 par. 1 EC,*** *since according to the case-law of the Court of Justice (Case C-53/00 [2001] Ferring ECR I-9067, paragraph 27), it was intended as compensation for the obligations of* ***services in the general economic interest*** *('SGEI obligations') imposed on all insurers active on the Irish PMI market*[362]*, namely* ***obligations designed to ensure that all persons living in Ireland would receive a minimum level of PMI services at an affordable price*** *and on* ***similar quality conditions.*** *The Commission observed (at recitals 40 and 41 to the contested decision) that that objective* ***would be achieved by establishing solidarity between policy-holders,***

[...]

166. As regards ***competence to determine the nature and scope of an SGEI mission within the meaning of the Treaty, and also the degree of control*** *that the Community institutions must exercise in that context, it follows from paragraph 22 of the Communication on SGEIs (see paragraph 12 above) and from the case-law of the Court of First Instance that* ***Member States have a wide discretion to define what they regard as SGEIs and that the definition of such services by a Member State can be questioned by the Commission only in the event of manifest error*** *(see case T-17/02 Olsen v Commission [2005] II-2031, paragraph 216 and the case-law there cited). [...]*

167. [...] It follows that the determination of SGEI obligations in this context also falls primarily within the competence of the Member States. *That* ***division of powers is also reflected, generally, in art. 16 EC,*** *which provides that, given the place occupied by SGEIs* ***in the shared values of the Union*** *as well as their role in promoting social and territorial cohesion, the Community and the Member States, each within their respective powers and within the scope of application of the Treaty, are to take care that such services operate on the basis of principles and conditions which enable them to fulfil their missions.*

[...]

276. In that regard, the Court considers that the theory that a ***danger of active risk selection*** *and, accordingly,* ***a risk of market instability is real and strong when the PMI insurer with a favourable risk profile embarks upon a strategy of price 'predation',*** *as described at paragraph 6 of the RES guide, is plausible. On the other hand, that danger appears to be less strong, but* ***still sufficiently significant,*** *when that insurer adopts— as the defendant, Ireland and the VHI claim with respect to BUPA Ireland's conduct—a* ***'price follower' strategy,*** *consisting in fixing the rates for its own PMI services at just below the rates charged by its main competitors for the same*

[362] Private medical insurance. Emphasis added by author.

services. In those circumstances, the customer would appear to have less financial incentive to change PMI insurers than in the case of 'price predation'. However, in that regard, the defendant and Ireland assert, without being contradicted by the applicants, that such a strategy none the less ***allows an insurer to select the 'good' risks*** *and that, especially, it affects the* ***proper functioning of the PMI market*** *and runs counter to the objective of community rating by maintaining premiums that are too high, to the detriment of customers, by comparison with the costs of claims which a PMI insurer with a favourable risk profile must actually bear.*

277. It also seems plausible that the 'price follower' strategy will have the effect that, notwithstanding its lower costs, the PMI insurer will decline to reduce its premiums, which would none the less be in the interest of consumers and consistent with the objective of community rating, with the sole aim of increasing its profits. Likewise, it does not appear to be precluded that a ***PMI insurer with an advantageous risk profile will pursue active risk selection by adopting the 'price follower' strategy so long as the price differential is sufficiently large to attract 'good risk' consumers.***

278. The Court considers that both Ireland and the Commission, at recital 50 to the contested decision, submitted plausibly that ***such a commercial practice was likely to create a downward spiral and thus to destabilise the equilibrium and the functioning of the community rating PMI market in so far as PMI insurers with a favourable risk profile would be increasingly able to attract young, healthy customers and thus further improve their risk profile, whereas PMI insurers with an unfavourable risk profile would come under growing financial pressure owing to the increasing disequilibrium between premiums and the costs occasioned by the claims submitted by a large number of elderly insured persons in poor health."***

The ECFI concluded that this was *not* a case of (prohibited) aid within the meaning of art. 87 TEC or new art. 107 TFEU but instead a permissible compensation system within the framework of a mission oriented towards serving the common good.

The judgment confirms the following:

- The understanding of what constitutes services rendered with an orientation towards serving the common good is the same under the law governing State aid and the public service clause of art. 106 par. 2 TFEU [86 par. 2].
- It is within the Member States' power of decision to determine what they consider such a service to be. As shown above, the Member States' autonomy concerning this matter has even been strengthened in the course of Lisbon by the new art. 14 TFEU which supersedes the previous art. 16 TEC.
- An insurance system that forcibly asserts solidarity, thus balancing good and bad risks and in particular preventing "cherry-picking" by individual insurers, represents a permissible service rendered with an orientation towards serving the

common good and not a (prohibited) aid scheme. Without such a configuration, the system would lose its financial equilibrium and the task oriented towards serving the common good could no longer be fulfilled.

C. Conclusion

The ECFI's justification can be applied to a statutory insurance system such as that of the KGV. Even within EU law governing State aid, it would be up to the Member States to determine whether a statutory insurance against natural hazards represents a service that is oriented towards the common good. Even the statutory insurance system according to the KGV example is dependent on solidarity and, for this reason, prevents "cherry-picking" regarding the good risks. Thus, it is also compatible with EU law governing State aid.

8. Outlook—Protection against natural disasters as a new EU goal

As a result of the Treaty of Lisbon revision, a new title XXII, Civil Protection, is to be found in the TFEU, with the following provision (emphasis added by the author):

"Title XXIII

Civil protection

Article 196

(1) The Union shall encourage cooperation between Member States in order to improve the effectiveness of ***systems for preventing and protecting against natural*** *or man-made* ***disasters.***

Union action shall aim to:

a. support and complement Member States' action at national, regional and local level in ***risk prevention****, in preparing their civil-protection personnel and in responding to natural or man-made disasters within the Union;*

b. promote swift, effective operational cooperation within the Union between national civil-protection services;

c. promote consistency in international civil-protection work.

(2) The European Parliament and the Council, acting in accordance with the ordinary legislative procedure shall establish the measures necessary to help achieve the objectives referred to in paragraph 1, ***excluding any harmonisation of the laws and regulations of the Member States.****"*

This new provision has already been discussed under the headings of public interest and harmonisation (see p. 116 et seq. above). Efficient systems

- to prevent natural disasters
- to protect against natural disasters
- to prevent risks

have in any case been incorporated in the new EU objective, not least in reaction to climate change.

This objective should be achieved without any harmonisation of the regulations of the Member States. It is thus left to the Member States as to how, and by what means, they choose to achieve this objective. The EU should merely facilitate such measures.

The future will show what effects this regulation will have in concrete terms. In any case, the KGV statutory insurance system against natural hazards, which also integrates the prevention of damage caused by natural hazards, is in accordance with the objectives of the new EU Treaties of Lisbon. It is not merely a system which is compatible with EU law but—measured against EU goals—is a system with the future in mind.

9. Final conclusion

A statutory insurance system against natural hazards, such as that of the KGV, is compatible with EU law.

EU law contains numerous foundations that can shore up this system. It thereby becomes apparent that the EU does not just permit such a system merely as an exception, but that it promotes them, particularly under the headings of general good, general interest services, services of general economic interest and civil protection.

A change of paradigm has taken place in the EU in the last ten years, which is now clearly reflected in the EU Treaties of Lisbon. Institutions and tasks that are oriented towards serving the common good now have a central place in the EU legal system. It is the responsibility of the Member States to decide whether a certain task, such as insurance against natural hazards, should be provided by a universal service system that is oriented towards serving the common good. EU entities must respect such a decision, and EU law contains several foundations on which a system to that effect may be legally incorporated, including at EU level.

As shown above, a statutory insurance system has numerous advantages. It is necessary in the face of climate change in order to guarantee comprehensive, across-the-board and affordable insurance protection to all building owners. Since a statutory insurance system such as that of the KGV is compatible with EU law, it represents an example worth emulating for the EU Member States as well.

Index

A

B

C

D

E

F

G

K

L

M

N

O

P

R

S

T

U

V

W

Z

MIX
Papier aus verantwortungsvollen Quellen
Paper from responsible sources
FSC® C105338

If you have any concerns about our products,
you can contact us on
ProductSafety@springernature.com

In case Publisher is established outside the EU,
the EU authorized representative is:
Springer Nature Customer Service Center GmbH
Europaplatz 3, 69115 Heidelberg, Germany

Printed by Libri Plureos GmbH
in Hamburg, Germany